Contents

Embodied Cognition and Shakespeare's Theatre

This collection considers issues that have emerged in Early Modern Studies in the past fifteen years relating to understandings of mind and body in Shakespeare's world. Informed by *The Body in Parts,* the essays in this book respond also to the notion of an early modern 'body-mind' in which Shakespeare and his contemporaries are understood in terms of bodily parts and cognitive processes. What might the impact of such understandings be on our picture of Shakespeare's theatre or on our histories of the early modern period, broadly speaking? This book provides a wide range of approaches to this challenge, covering histories of cognition, studies of early modern stage practices, textual studies, and historical phenomenology, as well as new cultural histories by some of the key proponents of this approach at the present time. Because of the breadth of material covered, full weight is given to issues that are hotly debated at the present time within Shakespeare Studies: presentist scholarship is presented alongside more historically-focused studies, for example, and phenomenological studies of material culture are included along with close readings of texts. What the contributors have in common is a refusal to read the work of Shakespeare and his contemporaries either psychologically or materially; instead, these essays address a willingness to study early modern phenomena (like the Elizabethan stage) as manifesting an early modern belief in the embodiment of cognition.

Laurie Johnson is Associate Professor of English and Cultural Studies at the University of Southern Queensland, Australia.

John Sutton is Professor of Cognitive Science, Macquarie University, Australia.

Evelyn Tribble is Donald Collie Chair of English at the University of Otago, Dunedin, New Zealand

Routledge Studies in Shakespeare

1 **Shakespeare and Philosophy**
Stanley Stewart

2 **Re-playing Shakespeare in Asia**
Edited by Poonam Trivedi and Minami Ryuta

3 **Crossing Gender in Shakespeare**
Feminist Psychoanalysis and the Difference Within
James W. Stone

4 **Shakespeare, Trauma and Contemporary Performance**
Catherine Silverstone

5 **Shakespeare, the Bible, and the Form of the Book**
Contested Scriptures
Travis DeCook and Alan Galey

6 **Radical Shakespeare**
Politics and Stagecraft in the Early Career
Christopher Fitter

7 **Retheorizing Shakespeare through Presentist Readings**
James O'Rourke

8 **Memory in Shakespeare's Histories**
Stages of Forgetting in Early Modern England
Jonathan Baldo

9 **Reading Shakespeare through Philosophy**
Peter Kishore Saval

10 **Embodied Cognition in Shakespeare's Theatre**
The Early Modern Body-Mind
Edited by Laurie Johnson, John Sutton, and Evelyn Tribble

Embodied Cognition and Shakespeare's Theatre

The Early Modern Body-Mind

Edited by Laurie Johnson, John Sutton, and Evelyn Tribble

Routledge
Taylor & Francis Group
NEW YORK AND LONDON

First published 2014
by Routledge
605 Third Avenue, New York, NY 10017

and by Routledge
2 Park Square, Milton Park, Abingdon, Oxon OX14 4RN

First issued in paperback 2021

Routledge is an imprint of the Taylor & Francis Group, an informa business

Library of Congress Cataloging-in-Publication Data

Embodied Cognition in Shakespeare's Theatre : the Early Modern Body-Mind / edited by Laurie Johnson, John Sutton and Evelyn Tribble.
pages cm. — (Routledge Studies in Shakespeare ; 10)
Includes bibliographical references and index.
1. English drama—Early modern and Elizabethan, 1500–1600—History and criticism. 2. Mind and body in literature. 3. Cognition and culture—England—History—16th century. I. Johnson, Lawrence, 1967– editor of compilation. II. Sutton, John, 1965– editor of compilation. III. Tribble, Evelyn B., editor of compilation.
PR653.E68 2014
822'.309353—dc23
2013041322

ISBN 13: 978-1-03-224292-7 (pbk)
ISBN 13: 978-1-138-00075-9 (hbk)

DOI: 10.4324/9780203796160

Typeset in Sabon
by IBT Global.

Figures

Acknowledgments

This collection of essays was first conceived on the back of a seminar at the 9th World Shakespeare Congress in Prague, 2011. The editors wish to thank all who participated in this seminar—the level of interest shown in the seminar by participants and auditors, the vibrant exchange of ideas that took place in this forum, and the collegiality and respect shown by all to each other convinced us that a book project on the theme of the early modern body-mind was achievable. The editors also thank the executive of the International Shakespeare Association—organizers of the Congress—for supporting the seminar and being freely ready with advice on how to successfully lead a large seminar of this kind. Thanks also go to those who have come on board with the project in the expansive life it has enjoyed since the decision was made to pursue the book in its current form.

The editors are indebted to their institutions—Macquarie University, the University of Otago, and the University of Southern Queensland—for material support in the development of this collection; particular gratitude is due to Professor Chris Lee, former director of the Public Memory Research Centre at the University of Southern Queensland, for ensuring that the funding and support necessary for the editing of this collection was available throughout the life of the project.

We also, each in our own way, extend our heartfelt appreciation to colleagues, friends, and family, who continue to fill our lives and work with rich rewards.

Introduction

Re-cognising the Body-Mind in Shakespeare's Theatre

Laurie Johnson, John Sutton, and Evelyn Tribble

The phrase "the mind-body problem" does not point to a single, unitary, perennial, and obvious human concern. Many people in different times and places have individually and collectively puzzled or agonized—in a range of intellectual, spiritual, and practical contexts—over the relations between various aspects of their nature which can operate in harmony or in tension. In English and other European languages, terms like *psychological* and *physical* have come to label what are sometimes seen as two realms or two sets of features and processes—the ingredients for that "mind-body problem." Yet both *body* and *mind* have complex and uncertain semantics that exceed the simple binary encapsulated within the parameters of this conceptual "problem." There is dramatic historical change and cross-cultural variation in the usage and meaning of *mind*, *psychology*, and *body*, of apparently central related general terms such as *cognition* and *consciousness*, and of many more specific "psychological" terms such as *emotion* and *memory*.[1]

By adopting the less familiar conjoined phrase *body-mind* in this volume, we seek therefore to defamiliarize our topics and to embrace the cultural, historical, and indeed scientific diversity of views, practices, and problems about thinking and the passions, imagining and dreaming, planning and communicating—about touch and vision and pain and fury. The essays we include cover an extraordinary array of "body-mind" topics, which cannot be reduced to singular terms. But even the label *body-mind*, of course, bears traces of the two connected dichotomous assumptions that our contributors seek to combat: the ideas that *mind* and *body* each name a unified set of phenomena held together by unique properties, and that there is thus a single problem about how they relate or connect. As a number of these essays suggest, we are so culturally marked by these historically specific assumptions that it is difficult to bracket them in addressing other ways of feeling, reasoning, remembering, or grieving embedded in quite different lived worlds.

In invoking the phrase *body-mind*, we also query the standard historical attribution of a damaging dualism to the "wound inflicted by the Cartesian split of mind and body."[2] Terms such as *pre-Cartesian* can function as convenient shorthand to note the characteristic melding of physiology and psychology so often seen in early modern humoral and medical discourse, as indeed it is used by many of our own contributors. But within the context of a volume of essays about the body-mind in the late sixteenth and early seventeenth centuries, we suggest that the labels *Cartesian*, *pre-Cartesian*, and *post-Cartesian* can mislead as well as inform our views of the body-mind. First, there were of course powerful earlier dualist frameworks in play in western culture: certain forms of Platonist and Christian thought and practice, to take just two key traditions, imposed or recommended dramatic divides between body and soul. Second, we risk a teleological reading of the unique historical discourses and feelings about the body around 1600 if we see these only as "pre-" Cartesian. We oversimplify if we simply read Descartes back into earlier and more alien nondualist frameworks for inhabiting the body-mind. Casting holistic humoral materialism as simple antecedents to a brutal "Cartesian" rupture or identifying putative early seventeenth-century signs of interiority and individualistic depth as mere harbingers of the Cartesian "invention of the mind" flattens out the complex landscape of the body-mind as it is diversely articulated in the early seventeenth century. Moreover, recent scholarship on Descartes's own work and context overturns the easy narrative by which the *cogito* alone was meant to ground an entire system of knowledge. Descartes was a natural philosopher of matter and motion who attributed an extraordinary range of capacities to biological systems embedded in complex environments; he also attributed memory, representation, sentience, and imagination to nonhuman animals. Descartes consistently focused in the human case on the union or integration of soul and body. Far from exclusively privileging the rational soul, his work substantially restricted its role and scope. He spent more time and energy working out our nature as complex bodies—not mere objects cut off from the world, responding passively to the whim of the soul—but fully and holistically embedded in the buzzing whirl of the fluid-filled cosmos.[3]

However we read the history of this debate, increasingly new research in the cognitive sciences and the humanities alike no longer univocally supports privileging the active rational mind over passive biology. We first might note that the cognitive sciences themselves are by no means a monolithic entity, instead denoting a diverse and often divided multidisciplinary field. To be sure, some dominant movements within it continue to model cognition as serial digital computation, or neurocentrically reduce thought and affect to brain processes alone. Classical cognitive science has been rightly criticized for its apparent *dis*embodiment, its assumption that the body and world primarily function as input-output devices for the brain, walling off perception and action from so-called internal cognitive processes, in what Susan Hurley has critiqued as the "classical sandwich."[4] However, in recent decades challenges to this model have become increasingly mainstream within the cognitive sciences. Cognition is increasingly seen, within these new frameworks, as "enactive," "embodied," "distributed," "situated," or "extended." Emerging paradigms in the cognitive sciences have increasingly sought to embody and extend cognition beyond the brain. The study of *situated* minds is intended to go decisively beyond what goes on in the individual skull, examining instead embodied, enactive, dynamic, and distributed cognitive processes as already bodily, social, practical, and worldly.[5] Thus the "mind" as it is currently conceived in many strands of the cognitive sciences is wildly heterogeneous, an on-the-fly assemblage of neural, kinesthetic, somatic, interpersonal, and material resources. We can, in other words, invoke these trends precisely to defamiliarize the idea of the separateness of bodies *and* minds.

Of course, researchers in the humanities long ago turned their attention to the body. Nearly two decades ago, the "bodily turn" was well enough established and diverse enough to provoke Caroline Bynum's question: "Why all the Fuss about the Body?"[6] As David Hillman and Carla Mazzio remark in *The Body in Parts*, in that essay Bynum had laid down the challenge for body work in the humanities since the body had by that time already become "no topic, or, perhaps, almost all topics."[7] The revisionary work carried out by Hillman and Mazzio and the contributors to *The Body in Parts* went far in both theorizing the early modern body and in carrying out an acute and particular examination of its "parts": entrails, nerves, breasts, bellies, brains, genitalia. Hillman and Mazzio caution against imposing a spurious unity upon the body: "in early modern representations, even or especially as a fantasy of the 'whole body' emerges, the body is at the same time always, and perhaps inevitably, a body in parts."[8] In drawing attention to the "materialist habits of early modern thought,"[9] Hillman in his contribution to *The Body in Parts* makes a strong case for what Gail Kern Paster terms the "early modern habits of bodily thought and sensation."[10] *The Body in Parts* consolidated earlier work and sparked sustained and productive attention to the body in the early modern period, especially within the context of Galenic thought and humoral theory, in which internal and external states are firmly yoked—indeed, mutually constitutive.[11] Moreover, research on the ways in which body itself is shaped by social and political forces has been a mainstay of New Historicist thought, particularly insofar as it has been influenced by the earlier work of Michel Foucault. Such work provides a rich foundation upon which the present collection of essays builds.

Yet as discussions of body waxed, explicit attention to mind waned, perhaps as a result of concerns that the concept paved the way for a latent universalism to emerge. In "Nervous Tension," Paster wrote of her desire to "produce a history of affects and behaviors resistant to the seductive evasions and erasures of essentialism."[12] If the body seems in danger of essentialist readings, perhaps even more so might the mind, particularly as it is sometimes construed in purely rationalist or "cognitivist" terms.[13] It must be admitted that there is a strongly universalist or essentialist strain in some so-called cognitive approaches to literature, especially as articulated in more extreme forms of literary Darwinism.[14] But as the authors of our essays show, attention to the concept of the "body-mind" can reinvigorate our understanding of the social and extended nature of cognition in the early modern period—what Paster describes in her linking piece in this collection as the "ecological framework of early modern personhood." In our view, expanding our ambit to the "body-mind" has the potential to open up questions of skill, animation, and kinesthesis. As Garrett Sullivan astutely observes in this volume, embodiment is sometimes viewed as a limiting boundary rather than as means of extension. Such concerns have been echoed in the emerging field of neuroanthropology, in which researchers such as Tim Ingold and Greg Downey have pointed out that the word *embodiment* can imply the subjugation of the body within a social field. As Ingold suggests, we might think "of the body not as a sink into which practices settle like sediment in a ditch, but rather as a dynamic center of unfolding activity."[15]

It is important to remember that these lines of inquiry are more than merely academic—it remains palpably evident that there is much at stake in the shaping of how Shakespeare and his contemporaries are viewed within the popular imaginary. A prelapsarian myth persists of Shakespeare and (as he tends to be too often set apart from) the early moderns. The discursive fields within which this version of history circulates extends beyond the reach of fiction and hyperbolic populism masquerading as criticism—Shakespeare's invention of the human, for example.[16] No less now than at any other time, the myths that persist within the broad popular imaginary also have their analogues within putatively objective medical or scientific writing. In two recent essays in medical journals, for example, medical practitioner Kenneth Heaton has argued that Shakespeare was particularly adept at describing the physical symptoms or sensory disturbances associated with emotional or mental upset.[17] By analyzing all of Shakespeare's plays and poems alongside forty-six genre-matched works by his contemporaries, Heaton showed that such conditions as breathlessness, fatigue, and vertigo were described far more often in Shakespeare's work in correlation with a heightened emotional state such as grief than in works by other writers of the same era, leading him to conclude, "Shakespeare's perception that numbness and enhanced sensation can have a psychological origin seems not to have been shared by his contemporaries."[18] Heaton's findings are *symptomatic*, we suggest, in two ways: first, they are symptomatic of the trend in popular appropriations of Shakespeare—they bear the hallmarks of modern readings that seek to dislocate the works of Shakespeare from their early modern moorings, to make Shakespeare a thoroughly modern thinker; second, Heaton's comments are "symptomatic" because they read surface phenomena as having "a psychological origin." According to this symptomatic reasoning, Heaton is prepared to read connections between sensations and emotions *in Shakespeare* as evidence of an awareness in the playwright that the origins of external distress reside in inner tumult, while at the same time he makes the mistake that Paster has consistently criticized scholars for making—of assuming that the psychological realm, for early moderns, is distinct from the physiological realm, and that thus physiological accounts of what we see as psychological phenomena must simply be metaphorical.[19] Yet, *non causa pro causa*—Shakespeare and his contemporaries did not view correlation as causation or metaphor in quite the same way as Heaton and his symptomatic reading.

Heaton dismisses in rather contemptuous fashion the medical knowledge of the early moderns:

> It is unsafe to try and contextualise the present findings by commenting on the occurrence of specific diseases in Shakespeare's time because concepts of physiology and disease were crude and fanciful, still dominated by Galenical ideas like the four humours. The circulation of the blood was not established until after Shakespeare died.[20]

All the more reason, we suggest, to not try to read Shakespeare and his coevals outside of their milieu—to resist the desire to want to recuperate Shakespeare as a timeless genius able to rise above the "crude and fanciful" notions that pervaded the thinking of all around him. No matter how crude or fanciful such notions may seem, the fact remains that the early moderns—Shakespeare included—left for us myriad disparate artefacts of the embodied cognition through which they conceived their world. One such artefact is the language itself—as the essays in *The Body in Parts* demonstrated, the language of the body was at one and the same time also the language through which abstractions were expressed, so a phrase like time being "out of joint" drew its particular force through reference to the painful physical dislocation of the joints.[21] Similarly, as Laurie Johnson has shown, the word *anxiety* comes into English at precisely this moment in history, circa 1611, because this is the moment when a need for such a word enters the language—what *anxiety* names is the very prospect confronting the early moderns that body and mind may be separable aspects of selfhood.[22] While Shakespeare never used the term, Johnson argues that he describes anxiety remarkably well, thereby contributing to the cultural moment out of which the word *anxiety* comes into being in its English form.

As we and our contributors articulate the "body-mind," then, it is open and contingent, a fluid assemblage that includes both a wealth of body-related parts and processes and a great diversity of psychological phenomena well beyond "subjectivity" and "mind," "self" and "inwardness." Our own interdisciplinary cognitive sciences expand to address material agency and embodied interaction, communication and collaboration, sociality and cognitive artefacts.[23] Likewise, the heterogeneous early modern body-mind phenomena discussed in the essays which follow range from wounds, torture, blood, scars, poison and blushing to guilt, sound, gesture, rhythm, sympathy, rage, madness, and anguish. Rather than seeking to divorce Shakespeare from his early modern milieu, the contributors to this collection consider the early modern theatre as offering fertile ground for reassessing the body-mind, whether through a distributed model of cognition for understanding early modern theatrical practices—as in the work of Evelyn Tribble[24] and in Johnson's account in this volume of the way in which this work guards against inherent limitations within the subversion-containment debates in Shakespeare Studies—or a reconsideration of the demonic infiltration of thought in the early modern period, as outlined here by Mary Floyd-Wilson; or the haptic nature of affect from the early modern period to the present, as Ros King, Jennifer Rae McDermott, and others explain in the present collection. In keeping with this sense that physical and psychological phenomena are complexly intercalated within the early modern body-mind, the editors have opted to avoid using any obvious internal divisions, enabling all of the chapters to speak to each of those that precede and follow them. Into this mix we have a number of shorter linking pieces that respond in a variety of ways to issues and arguments raised in the chapters surrounding them.

In the first chapter, David Hawkes examines early modern approaches to torture and physical pain as the litmus test for Francis Bacon's descriptions of the adversarial relation in which the empirical scientist tortures nature into giving up her secrets. Shakespeare's frequent deployments of the same image serve to defend the nonmaterial subject or "soul" from encroaching materialist thought. Shifting our attention into the playhouse, Ros King is nevertheless concerned in the second chapter with the extent to which mindfulness and the expression of feeling among performers and audience can be tested in the dynamic play of the playhouse. There is thus in the first two chapters a question mark placed at the limits of empiricism, and King reminds us that the experience of the play is a legitimate study of these limits, as Shakespeare's depiction of audience participation in *Much Ado About Nothing* bears witness.

While playing creates the interpersonal space in which meaning—and, by association, the expression of feelings—can be negotiated rather than simply communicated, Emma Firestone's analysis of the language of *hot* and *cold* in relation to Prince Hal suggests that the interplay of word and body-mind is possibly cognitively weighted in favor of the word. When Hal is cold and Hotspur is, well, hot, at least in the language through which we understand their qualities, our interpersonal inclinations follow a deep-seated bias in which warmth equals trustworthiness. The reader who peruses this book sequentially will encounter the first of our linking segments at this point, wherein Garrett Sullivan takes up the challenge of the first three chapters by looking to the work of Thomas Kyd and John Webster as examples of how, *pace* Shakespeare, there is a widespread concern within Renaissance drama with the performance of subjectivity. The subject that treads the boards of the early modern stage is no monad—it is actively worked upon by and works upon forces that shape the world. One such force—as Mary Floyd-Wilson explains in her reading of Thomas Heywood's *A Woman Killed with Kindness*—is diabolical indeed. It should be absurd to think of discussing early modern cognition without reference to the soul and the myriad temptations to which it is heir, yet Floyd-Wilson notes that readings of Heywood's play have tended to read Anne's adultery as a consequence of her inability to think for herself due to dominant gender roles in the marital arrangement. Floyd-Wilson argues compellingly instead that Anne gives in to temptation as a demonstration of "the period's most pressing question about cognition: Can you call your thoughts your own, or were they planted by the Devil?"

Michael Schoenfeldt's linking segment, which follows James A. Knapp's treatment of scenes of misrecognition in *Much Ado About Nothing*, pursues the pressing question posed by diabolical temptation as a subset of the more general doubt pervading the early modern period about the permeability of the body-mind. Knapp's discussion of misrecognition and the staging of theatrical illusion in *Much Ado* might be compared here with King's study of play within the same play. Again, the limits of perception are in question. Similar comparisons might be made between the chapters by Darryl Chalk and Hardin Aasand, whose respective readings of *The Winter's Tale* we position here side by side. Whereas heat might have rendered Hotspur more trustworthy by comparison with the cold Prince Hal in Firestone's reading of the Henriad, the potential is offered in *Winter's Tale* for a concatenation of forces to come together to make Leontes "too hot," which Chalk explains as a result of "contagion by image." The same is also presented by Chalk as an explanation for the seemingly inexplicable death of Mamillius. By contrast to Chalk's reading of the sudden ruptures plaguing the play's opening scenes, as Schoenfeldt explains, Aasand offers a reading in which the figure of Time, like the Devil, can function as a singularity upon which perception can be fixed as a constant external reference point for the unreliable human senses.

Paster's linking segment weaves together numerous threads of our collection through her reading of the ecology of the passions as it is played out in the trope of the spider in the cup in *The Winter's Tale*. This play has become a sort of touchstone in our collection, a singularity fixing our critical gaze. Perhaps it is in Shakespeare's late plays, and the genre of romance, that the imbrication of the "body-mind" becomes most visible. Jennifer Rae McDermott focuses our attention further on the spider's touch in *Othello*, but she does so in order to point out that the metaphor of spider's touch-in-web functions both on an individual level, in the desire to hold woven properties, and on a communal level, where threads forge networks of feeling among the audience. Tiffany Hoffman's reading of the blush in *Coriolanus* likewise emphasizes the social dimension of perception, wherein the surface appearance is meaningless without the observer to participate in apprehending its meaning. A tension may therefore be seen to be emerging here: whereas earlier chapters deal with the potential for the interpersonal relation to be shaped by relatively fixed universes of meaning around specific words and phenomena, these later chapters identify the confluence of the interpersonal as the site of dynamic cognitive processes.

Katherine Rowe provides the link toward a model of distributed cognition that emerges logically in this last grouping of chapters. As Rowe also mentions and as Johnson argues in his chapter, the distributed cognition model finds a natural home within cultural histories of early modern cognition, given the importance placed in both on the constitutive role of the material artefact in dynamic *cultural* processes. Lianne Habinek's chapter does not focus on distributed cognition per se but it does, importantly, situate early modern thought about the brain and brain injury within a cultural history of early modern medical discourses, which, in turn, is depicted in Hamlet's concern with the impermanence of the "book and volume" of his own brain. Heaton may dismiss early modern medical beliefs as crude and fanciful, yet Habinek maps the broad cultural field in which these early modern beliefs pave the way for modern lesion studies. Jan Purnis maps a similar cultural field around the early modern belief in the localization of the "second brain" in the human gut and digestive system, which modern studies into the enteric nervous system are showing to also be far from crude or fanciful. For both Habinek and Purnis, then, a particular body part is seen as participating in the early modern understanding of mind and mental phenomena, yet these bodily parts are also fundamentally porous, as we have seen, and are capable of being imprinted and reprinted anew on the basis of external phenomena. In such an arrangement, cognition does not happen in response to the outside world; rather, there is constant interplay between bodies wherein genuinely embodied cognition occurs. Every chapter in this volume speaks in some way, and from very different backgrounds and approaches, to this arrangement that we seek to understand on this side of history through the invocation of the "body-mind"—which we use here devoid of any qualifying phrase: *sans* "relationship," *sans* "dualism," and certainly *sans* "problem."

NOTES

1. For discussions of these issues, see Anna Wierzbicka, *Semantics, Culture, and Cognition: Universal Human Concepts in Culture-Specific Configurations* (Oxford: Oxford University Press, 1995); Graham Richards, *Mental Machinery: The Origins and Consequences of Psychological Ideas Part 1: 1600 to 1850* (London: Athlone Press, 1992); Timothy J. Reiss, "Denying Body, Making Self?: Histories and Identities," in *Against Autonomy: Global Dialectics of Cultural Exchange*, ed. Timothy J. Reiss (Stanford: Stanford University Press, 2002): 184–218; Paul S. Macdonald, *History of the Concept of Mind: Speculations about Soul, Mind, and Spirit from Homer to Hume* (Aldershot: Ashgate, 2003); Mengistu Amberber, ed., *The Language of Memory in a Cross-Linguistic Perspective* (Amsterdam: John Benjamins, 2007); W. I. Matson, "Why Isn't the Mind-Body Problem Ancient?," in *Mind, Matter, and Method*, ed. Paul K. Feyerabend and Grover Maxwell (Minneapolis: University of Minnesota Press, 1966): 92–102; and Peter King, "Why Isn't the Mind-Body Problem Medieval?," in *Forming the Mind: Essays on the Internal Senses and the Mind-Body Problem from Avicenna to the Enlightenment,* ed. Henrik Lagerlund (Berlin: Springer, 2007): 187–205.
2. Maxine Sheets-Johnstone, "Animation: The Fundamental, Essential, and Properly Descriptive Concept," *Continental Philosophy Review* 42.3 (2009): 375–400; esp. 375.
3. See Stephen Gaukroger, *Descartes: An Intellectual Biography* (Oxford: Clarendon Press, 1995); John Sutton, *Philosophy and Memory Traces: Descartes to Connectionism* (Cambridge: Cambridge University Press, 1998); Catherine Wilson, "Descartes and the Corporeal Mind: Some Implications of the Regius Affair," *Descartes' Natural Philosophy*, ed. Stephen Gaukroger, John Schuster, and John Sutton (London: Routledge, 2000): 659–679; and Reiss. As Gaukroger notes, for at least a century after Descartes's death, his philosophy was attacked less for its incoherent dualism than for the specter of atheistic materialism that both conservatives and enthusiasts read into it: with his scientific or natural-philosophical works read more than the *Meditations,* it was arguably only after Kantian historians of philosophy identified Descartes as the arch-rationalist that he came to play the familiar textbook role of the evil demon at the origin of the modern world.
4. Susan Hurley, "Perception and Action: Alternative Views," *Synthese* 129.1 (2001): 3–40.
5. The most important single source for these movements is Andy Clark, *Being There: Putting Brain, Body, and Mind Together Again* (Cambridge, Massachusetts: Massachusetts Institute of Technology Press, 1997). For broader takes on their history and current status, see Margaret Boden, *Mind as Machine: A History of Cognitive Science* (Oxford: Oxford University Press, 2006); Philip Robbins and Murat Ayede, *The Cambridge Handbook of Situated Cognition* (Cambridge: Cambridge University Press, 2009).
6. Caroline Bynum, "Why all the Fuss about the Body: A Medievalist's Perspective," *Critical Inquiry* 22 (Autumn 1995): 1–33.

7. Bynum, 2; see David Hillman and Carla Mazzio, *The Body in Parts: Fantasies of Corporeality in Early Modern Europe* (New York and London: Routledge, 1997), xii.
8. Hillman and Mazzio, xviii.
9. David Hillman, "Visceral Knowledge," in Hillman and Mazzio, 81–106, esp. 82.
10. Gail Kern Paster, "Nervous Tension," in Hillman and Mazzio, 107–127, esp. 107.
11. See, among others, Michael Schoenfeldt, *Bodies and Selves in Early Modern England: Physiology and Inwardness in Spenser, Shakespeare, Herbert, and Milton* (Cambridge: Cambridge University Press, 1999); Gail Kern Paster, *Humoring the Body: Emotions and the Shakespearean Stage* (Chicago: University of Chicago Press, 2004); and for an overview, see Sean McDowell, "The View from the Interior: The New Body Scholarship in Renaissance/Early Modern Studies," *Literature Compass* 3:4 (July 2006): 779–791; Mary Floyd-Wilson and Garrett Sullivan, eds., *Environment and Embodiment in Early Modern England* (London: Palgrave Macmillan, 2007); and Laurie Johnson, "Differing Returns: On History, Bodies and Early Modern Lives," *Early Modern Literary Studies* Special Issue 19 (2009): 14.1–9.
12. Paster, "Nervous Tension," 123.
13. Slavoj Žižek, for instance, equates "cognitivism" with rational managerialism; see "Cultural Studies versus the 'Third Culture,'" *The South Atlantic Quarterly* 101.1 (2002): 19–32.
14. For a critique of such models, see Jonathan Kramnick, "Against Literary Darwinism," *Critical Inquiry* 37 (2011): 315–347. For alternatives to this universalist strain, see, for example, John Sutton and Evelyn Tribble, "Minds In and Out of Time: Memory, Embodied Skill, Anachronism, and Performance," *Textual Practice* 26.4 (2012): 587–607; and Mark J. Bruhn, "Introduction: Exchange Values: Poetics and Cognitive Science," *Poetics Today* 32.3 (2012): 403–460.
15. See Greg Downey, "Balancing between Cultures: Equilibrium in Caporeia," in *The Encultured Brain: An Introduction to Neuroanthropology*, ed. Daniel Lende and Greg Downey (Cambridge, Massachusetts: Massachusetts Institute of Technology Press, 2012): 169–194, esp. 187; Tim Ingold, "Towards an Ecology of Materials," *Annual Review of Anthropology* 41 (2012): 427–442; esp. 437.
16. Harold Bloom, *Shakespeare: The Invention of the Human* (New York: Fourth Estate, 1998).
17. Kenneth Heaton, "Body-Conscious Shakespeare: Sensory Disturbances in Troubled Characters," *Medical Humanities* 37.2 (2011): 97–102; "Somatic Expressions of Grief and Psychosomatic Illness in the Work of William Shakespeare and His Coevals," *Journal of Psychosomatic Research* 73.4 (2012): 301–306.
18. Heaton, "Body-Conscious," 101.
19. Gail Kern Paster, *The Body Embarrassed: Drama and the Disciplines of Shame in Early Modern England* (Ithaca, New York: Cornell University Press, 1993).
20. Heaton, "Somatic," 304.
21. Marjorie Garber, "Out of Joint," in Hillman and Mazzio, 23–52.
22. Laurence Johnson, "'Nobler in the Mind': The Emergence of Early Modern Anxiety," *AUMLA* Special Issue (December 2009): 141–156.
23. Edwin Hutchins, "Material Anchors for Conceptual Blends," *Journal of Pragmatics* 37 (2005): 1555–1577; Lambros Malafouris and Colin Renfrew,

eds., *The Cognitive Life of Things* (Cambridge: McDonald Institute, 2010); Jürgen Streeck, Charles Goodwin, and Curtis LeBaron, eds., *Embodied Interaction: Language and Body in the Material World* (Cambridge: Cambridge University Press, 2011).

24. Evelyn B. Tribble, *Cognition in the Globe: Attention and Memory in Shakespeare's Theatre* (New York: Palgrave Macmillan, 2011).

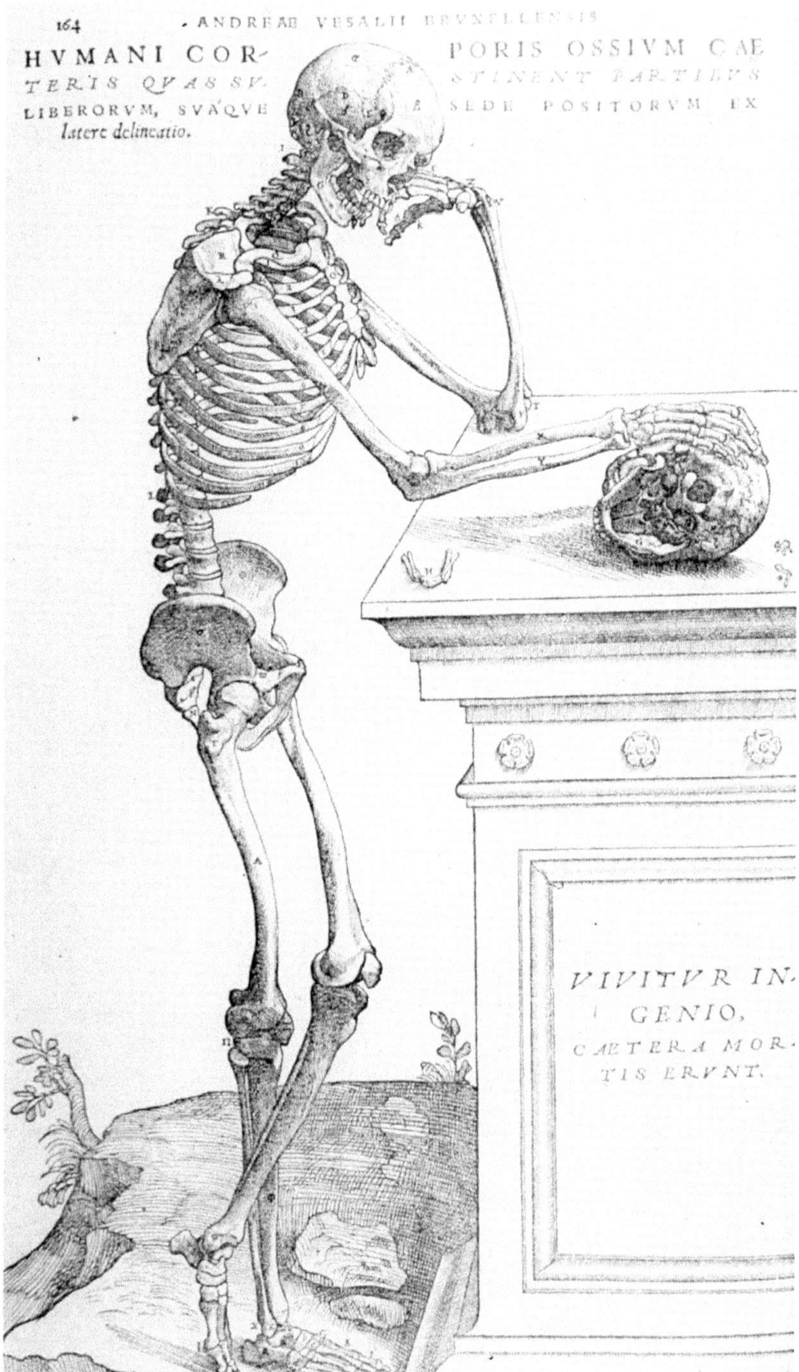

Figure 1 “Bone Plate,” from Vesalius, *De humani corporis fabrica* (1543), 164 (Liber I). Image A026932, History of Medicine Collection, Courtesy U.S. National Library of Medicine.

1 Proteus Agonistes

Shakespeare, Bacon, and the "Torture" of Nature

David Hawkes

Bacon seems to have written his essays with Shakespeare's pen.

—Alexander Smith[1]

It has been said, that in the constructing of Shakespeare's Dramas there is, apart from all other "faculties' as they are called, an understanding manifested, equal to that in Bacon's *Novum Organum*. That is true; and it is not a truth that strikes every one.

—Thomas Carlyle[2]

the philosophical writings of Bacon are suffused and saturated with Shakespeare's Thought.

—Gerald Massey[3]

PART I

As these quotations from eminent Victorians suggest, the possibility of a mutual influence between Bacon and Shakespeare has intrigued readers for nearly two centuries. Generations of critics have remarked on the extraordinary affinity between the two men's thought, and it has often been suggested that Shakespeare gives poetic, imaginative expression to the same ideas that Bacon advances in prosaic, discursive form. So striking is this resemblance that some have even been tempted to assert that Bacon is the true author of the works attributed to Shakespeare. The evidence alleged in support of this case is sometimes eccentric, but it is not hard to believe that the Zeitgeist, which generally inspires many individuals simultaneously, affected the minds of Bacon and Shakespeare in a broadly similar manner.

In what follows I examine Shakespeare's treatment of an especially important Baconian metaphor and ask whether the two writers' deployment of this image might reveal a deeper set of conceptual connections between them. Notoriously, Bacon compares the empirical method of investigating nature to torture. This trope has considerable implications for both the human mind and objects of its study. Bacon has frequently been criticized for suggesting that "art" and "nature" exist in a basically antagonistic relation, to the degree that art must (and should) torture nature into divulging her secrets: "the nature of things reveals itself more readily under the vexations of art than in its natural freedom."[4] The purpose of "art," by which Bacon means empirical investigation by controlled experiment, is above all instrumental. Art seeks to exploit nature for the ends of the artist and, as with analogous forms of exploitation, this involves the use of violent coercion. Proteus, Bacon's figure for matter, must be forcibly constrained into lending his assistance to human beings: "if anyone wanted his help in any matter, the only way was first to secure his hands with handcuffs, and then to bind him with chains."[5] The heroic scientist who wrestles with Proteus is one who "shall bring force to bear on matter and shall vex it and drive it to extremities" (6.726).

This image of science as "torturing" Proteus, or nature, is among the most controversial tropes in the history of ideas. Descriptions of nature as Protean had been commonplace since ancient times, and the study of nature was conventionally referred to as "wrestling with Proteus."[6] In *The Garden of Cyrus,* Sir Thomas Browne calls Proteus a "Symbole of the first matter,"[7] the formless *prima materia* before God created nature by His art. This symbolic Proteus designates nature, viewed from the standpoint of art, as an unpredictable, infinitely malleable force in need of discipline and control if its power was to be directed toward the good. In fact, Proteus was a convenient figure for ambiguity and shape shifting in general. In addition to representing the empirical investigator's view of nature, he was a conventional symbol for verbal equivocation. The *Treatise of Equivocation,* discovered in the possession of the Jesuit Father Garnett, warns against the exploitation of ambiguity by opportunists "more inconstant than *Proteus*."[8] This association forged another nexus between Proteus and torture, since the controversy over the Jesuits' endorsement of equivocation directed attention to the issue of whether torture can ever provide access to the truths of the psyche.

Bacon's contemporaries reacted with suspicion to his aggressive rhetoric, and their criticisms have echoed down the centuries. For the Romantics, Baconian empiricism was not an aid but a hindrance to authentic knowledge. Goethe protested that "Nature falls silent on the rack."[9] In recent decades, feminists, ecologists, postcolonialists, and Frankfurt School Marxists have presented different yet equally cogent critiques of Bacon.[10] In his most villainous incarnation, he is presented as instigating a theory of knowledge as an intrusive violation of nature by art. According to Carolyn Merchant, his method was modeled on "the interrogations of the witch trials," so that "the very essence of the experimental method arose out of human torture transferred onto nature."[11] Elizabeth Hanson identifies Bacon's hypothetical ideal scientist "as a champion of the discovery of nature's secrets and as a persistent practitioner of torture."[12] By the end of the twentieth century, Bacon's position as "the whipping boy of ecological politics" was firmly established.[13]

However, a significant tendency in recent scholarship, exemplified by Peter Pesic's fascinating series of essays "Wrestling with Proteus," "Proteus Unbound," and "Proteus Rebound,"[14] suggests that Bacon's metaphor has been badly misunderstood. Pesic argues that Bacon actually envisaged a complementary relation between art and nature, in which they cooperate harmoniously in the revelation and direction of natural forces. By using the noble, dramatic figure of Proteus as his symbol for nature, claims Pesic, Bacon portrays the encounter of art and nature as "a heroic mutual struggle, not the torture of a slavish victim. By the 'vexation' of nature Bacon meant an encounter between the scientist and nature in which both are tested and purified."[15] Despite Pesic's revisionism, however, it is undeniable that the period's literature often attacks the experimental attitude to nature as destructive and exploitative. If this suspicion was not generally directed at Bacon personally, that suggests that it was too broad and pervasive to concentrate on a single individual. One of the points I try to make here is that Shakespeare was also meditating on torture and truth—at the same time, in the same place, and to many of the same ends as Bacon.

Pesic cites several seventeenth-century commentators, including Samuel Hartlib and John Milton, to show that they did not interpret Bacon as advocating a hostile approach to nature. With commendably full disclosure, however, he cites passages where even his exemplary authors appear to condemn the empirical method. Hartlib, for example, considers it "impiety" to "mangle tyrannise etc. over the Creatures for to trie exp[e]riments."[16] I have discussed elsewhere how, in Milton's *Paradise Lost,* the archangel Raphael cautions Adam against the misapprehensions that arise from empirical investigation of the environment.[17] Adam finds it strange that God has ordered the universe in such a way that greater planets like the sun exist only to serve smaller orbs such as the earth. Raphael, who is studying Adam as closely as the human is studying him, immediately understands that his instinct to evaluate his surroundings on the basis of sense data reveals a naturally servile tendency of mind which, he correctly predicts, will cause human history to be dominated by political tyranny: "Already by thy reasoning this I guess/ Who art to lead thy offspring." (8:85–86)

Raphael warns Adam to seek out the nonapparent essences behind appearances, reminding him that "great/ Or bright infers not excellence" (8:90–91), but Adam's subsequent confessions concerning his sexuality suggests that this caution must fall on deaf ears. The puny mortal once again dares to criticize his Creator, this time on the grounds that He has made Eve so physically beautiful as to obviate man's use of reason. He blames her body for his own aberrant thoughts, lamenting that human woe begins with female beauty and earning a stern rebuke from his new instructor, the archangel Michael: "From man's effeminate slackness it begins" (11.634). The problem does not lie with Eve's body and still less with the God who made it but with man's tendency to reify and fetishize his objects of desire, evaluating them on the basis of their sensual appearances alone. In spite of his admiration for Bacon, Milton presents empirical perception as necessarily misleading. *Paradise Lost* teaches that our tendency to mistake the phenomenal world for reality is both cause and effect of the Fall.

So the philosophy we know as "empiricism," and the instrumental study of nature that results, were regarded as distinctly dubious in early-seventeenth-century England. The very word *empiric* was commonplace shorthand for the shadier sort of alchemist, petty conjurer or hedge witch. The charge was frequently made that—like such primitive magic—empiricism is concerned with appearances to the exclusion of essences; that it concentrates on accidents and disregards substances; and that this entails a basic misconception as to the nature of reality itself. Bacon himself claimed not to care about the ethical implications of the terms in which his enterprise was described: "I intend and mean only that nature, like *Proteus,* is forced by art to do that which would not have been done without it: and it does not matter whether you call this forcing and enchaining, or assisting and perfecting." Subsequent commentators have vigorously taken issue with Bacon on this point, noting that the way we frame the relations between "art" and "nature" carries implications reaching far beyond Bacon's own sphere of experience.

If Pesic's analysis of the metaphor is correct, so that Proteus and the scientist work together for mutual benefit, many of today's criticisms of science's allegedly imperialist attitude toward nature would have to be reconsidered. Science would appear fundamentally ameliorative, and the supposed alienation of the human subject from its objective context would come to seem a less pressing moral problem. On the other hand, if Bacon conceived of his enterprise as aggressive and exploitative, even unconsciously, this would lend support to the ethical case against instrumental reason and empirical science. The rest of this chapter examines the ways in which Shakespeare takes up this debate in fictional terms. Does he envisage "art" and "nature" as existing in a hostile hierarchy, or does he consider them at least potentially harmonious and complementary?

PART II

The practical benefits of Bacon's method would soon become undeniable. Yet many of his contemporaries believed that the "new philosophy" perverted the respectful study of nature into a tool for its exploitation. As Gloucester, spokesman for the conservative world picture, laments in *King Lear,* "[t]hough the wisdom of Nature can reason it thus and thus, yet Nature finds itself scourged by the sequent effects" (1.2.106–108). The practical effect of natural philosophy—the "wisdom of nature"—is the torture of nature itself. In *The Tempest,* Miranda's opening lines reveal her suspicion that the shipwrecking storm was Prospero's doing: "If by thy art, my dearest father, you have/ Put the wild waters in this roar" (1.2.1–2). The benignity of his power is by no means obvious here. And yet in *The Winter's Tale*, Shakespeare portrays "art" as aiding or expediting natural processes toward their proper ends when Polixenes describes the grafting of flowers as "an art/ Which doth mend Nature—change it rather; but/ The art itself is Nature" (4.4.95–97).

On this issue Shakespeare displays what Keats called his "negative capability" of holding contradictory opinions simultaneously. The logic and morality of empirical science is debated throughout his work, and his noncommittal, perplexed attitude is typical of the first generation of Bacon's readership. Each of the three "last plays"—*Cymbeline, The Winter's Tale*, and *The Tempest*—focuses on a child of aristocratic genealogy but impoverished upbringing and asks whether nature or nurture determines its character. In *The Winter's Tale,* it is Perdita's artificial "unusual weeds" that make her natural breeding visible to Florizel, who instantly recognizes that she is "no shepherdess" (4.4.2–3), and "too noble for this place" (4.4.159). In *The Tempest,* Prospero's punning address to "my daughter, who/ Art ignorant of what thou art" (1.2.17–18) hints that nobility may be a matter of "art" or mere convention, but the ease with which Miranda falls in love with Ferdinand reveals her to be a natural aristocrat.

Prospero dominates the island's natural forces by controlling the labor of his slave Caliban and his servant Ariel. The pronounced difference between slave and servant parallels the distinction between an exploitative and a cooperative relation between art and nature. Prospero reminds Ariel that "it was mine art/ When I arrived and heard thee, that made gape/ The pine, and let thee out" (1.2.426–428). Far from inflicting torture on Ariel, Prospero's art liberates him from his previous "torments." Their cooperation, though uneasy and fraught at times, is basically beneficial to both parties. By contrast, Prospero's use of his "art" to enslave Caliban illustrates, in *Dialectic of Enlightenment*, Adorno and Horkheimer's description of Bacon's method: "What human beings seek to learn from nature is how to use it to dominate wholly both it and human beings" (2). As Ayanna Thompson and others have explained, slavery and torture have been close conceptual companions throughout the ages.[18] The pattern of association between carnality and servility that characterizes so much early modern literature stems from Aristotle's analogy between the master-slave relationship and the relation between the soul and the body: "that which can foresee with the mind is the naturally ruling and naturally mastering element, while that which can do these things with the body is the naturally ruled and slave."[19] Christianity takes up this imagery in the New Testament, especially the Pauline epistles: "I am carnal, sold under sin" (Romans 7.14). Carnality was both the result and the justification of slavery; the carnal man is a natural slave, and slavery is his proper position.

Caliban can be viewed as the clearest instance of a natural slave in Shakespeare. That is certainly how Prospero sees him. This master refers to his slave as "earth," as a "thing," and as one receptive only to physical stimuli: "[w]hom stripes may move, not Kindness" (1.2.496–497).[20] Physical torture was reserved for slaves in the ancient world, for only slaves were conceived as susceptible to such carnal coercion. Slavery itself was understood as a condition of objectification. Although his conversation shows that he does possess an interior subjectivity, Caliban is treated by others as if he were nothing but brute matter. As such, he is physically forced to follow the dictates of Prospero's "art": "I must obey, his art is of such power" (1.2.375–376). Caliban identifies this "art" with the capacity to inflict physical pain, informing Trinculo, "Thou dost me yet but little hurt; thou wilt anon,/ I know it by thy trembling: now Prosper works upon thee" (2.2.28–29).

The Winter's Tale also establishes and defines a servile character by his subjection to corporal punishment. Autolycus introduces the theme in his introductory scene: "gallows and knock are too powerful on the highway. Beating and hanging are terrors to me" (4.3.25). When he dons a nobleman's clothes and seeks to act with authority, he reveals his natural servility by the manner in which he imitates power. Told that the prince intends to marry a shepherd's daughter, he attempts to ape authority by angrily describing "the tortures he shall feel" (4.4.769). In a comic demonstration of his servile ineptitude, he becomes carried away, threatening to torture even the shepherd's son, who "shall be flay'd alive; then 'nointed over with honey, set on the head of a wasp's nest" (4.4.783–785). The clown's observation that "[h]e seems to be of great authority" (4.4.800) reveals the same, simple identification of authority with the power to torture.

An atmosphere of ethical suspicion, easily missed by modern readers, envelops Prospero's "art" from the beginning. He tells Miranda how he neglected his duties as duke of Milan and "to my state grew stranger, being transported/ And rapt in secret studies." The suggestion that he was studying magic is confirmed by his power on the island. Although he manufactures the tempest for the beneficent purpose of ameliorating the souls of its victims, official legal and theological doctrine acknowledged no distinction between "white" magic and *maleficia,* while the ability to control the weather by means of "art" was one of the powers regularly ascribed to witches. The weird sisters in *Macbeth* plot to disrupt the voyage of a ship, and by analogy the equanimity of Macbeth's soul, through the same means employed by Prospero: "Though his bark cannot be lost,/ Yet it shall be tempest-tost" (1.3.26–27). In *Macbeth* as in *Lear,* Shakespeare's imagery subjects nature to monstrous violations, as predators are devoured by prey and horses eat each other. "'Tis unnatural," as the Old Man observes, "[e]ven like the deed that's done" (2.4.11–12). During his final interrogation of the witches, Macbeth's quest for knowledge becomes openly destructive: "though the treasure/ Of nature's germens tumble all together,/ Even till destruction sicken; answer me/ To what I ask you" (4.1.58–61).

The usurpation of the throne is succinctly described by Lady Macbeth as "nature's mischief." It sets in motion a series of correspondingly unnatural occurrences, directed from outside by the forces of black magic and also penetrating deep within the minds of its perpetrators. Both Scotland as a whole and the psyches of the play's villains are subjected to torture. In the midst of his vexation, Macbeth cries out that he envies those he has killed, since he would rather die "[t]han on the torture of the mind to lie/ In restless ecstasy" (3.2.22–25). His words anticipate Elaine Scarry's account of torture's "obliteration of consciousness."[21] Macbeth's mind has become something other than himself; a thing on which he may "lie," but whose division from the self is signified by the term *ecstacy*: "standing outside." In the final act, this interior mind is finally brought, like the victim of physical torture, to denounce itself: "all that is within him does condemn/ Itself for being there" (5.2.28–29). In *Macbeth,* Shakespeare asks what effect the conception of "art" as the torture of nature might have on the soul of the "artist." What happens to the human subject which suddenly finds itself cast in the role of torturer?

PART III

Today, when the citizens of advanced democracies earnestly debate the ethical merits of literal, physical torture—in which their governments are vigorously engaged—the relation between the tortured body and the interior mind is once again a topical issue.[22] In the wake of September 11, 2001, a posse of commentators rushed to rationalize the torture of suspected terrorists. The most notorious argument was Alan Dershowitz's "ticking time bomb scenario," which called for the legalization of torture in advance of a situation where, it is claimed, it would have to be used: a prisoner who knows but will not reveal the location of a bomb certain to kill millions should it explode.[23] This implausible scenario is deemed fascinating and complex enough to be the subject of its own television series, named *24,* in which U.S. security agents commit various despicable atrocities in their search for truth. The thrill presumably experienced at this spectacle by sections of the public is not mere sadism. It is the comfort that arises from knowing with indubitable certainty, confirmed weekly in fictional but forceful terms on millions of television sets, that torture is efficacious. Indeed, so satisfying is this certainty that it has been invoked as demanding a thoroughgoing revision of America's political morality. Writing in *Newsweek* in November 2011, Jonathan Alter asserted that "[s]ome torture clearly works," and boldly drew the logical conclusions from this apparently undoubted efficacy:

> We can't legalize torture; it's contrary to American values. But even as we continue to speak out against human-rights abuses around

> the world, we need to keep an open mind about certain measures to fight terrorism, like court-sanctioned psychological interrogation. And we'll have to think about transferring some suspects to our less squeamish allies, even if that's hypocritical. Nobody said this was going to be pretty.[24]

The assumption that torture "works" was also widespread in the age of the witch trials. As is his habit, Shakespeare approaches the matter from several angles at once during the denouement of *Othello,* in which the stunned surviving characters debate the perplexing mystery of Iago's motive. In the process, they force us to consider the relation between internal motive and external act. As he has throughout the play, the villain protests his honesty: "I told him what I thought" (5.2.183). This statement is unverifiable from outside Iago's psyche. When Othello demands "[w]hy he hath thus ensnared my soul and body" (5.2.346), Iago vows that his appearance is all they will ever see: "Demand me nothing: what you know, you know:/ From this time forth I never will speak word" (5.2.347–348).

The truly frightening prospect raised by the play's conclusion is that Iago has no motive because he has no soul. He is hiding nothing; he actually *is* only appearance. The other characters initially determine to force the revelation of Iago's motive by torturing him. Gratiano confidently predicts that "[t]orments will ope your lips" (5.2.350), and Othello appears to agree: "Well, thou dost best" (5.2.351). There must be an interior psyche in there somewhere, they assume; the only problem is how they can force it to reveal itself. Yet following the discovery of letters revealing the mechanics of the plot, though not the motives of its designer, Lodovico appears to forget the forensic purpose of torture in a pure lust for vengeance: "If there be any cunning cruelty/ That can torment him much and hold him long,/ It shall be his." Lodovico's final appeal that the governor enforce "the censure of this hellish villain;/ The time, the place, the torture" (5.2.413–414), shows that he anticipates torture as part of the sentence rather than the investigation. It is an admission of defeat. The audience certainly never learns Iago's motive to any satisfying degree, and this draws our attention inexorably to the possibility that Iago is what he does and nothing more. His famous boast "I am not what I am" is thus doubly ironic: he is exactly what he seems to be, for there is no distinction between appearance and reality in his world. Torture may slake the thirst for vengeance, but it cannot reveal a man's ulterior motive if no such thing exists.

In *The Merchant of Venice,* Portia also denies that torture can reveal a man's authentic thoughts. The speech of a tortured man does not come from the heart; it is not sincere. As she tells Bassanio: "I fear you speak upon the rack,/ Where men enforced do speak anything" (3.2.1396). Under torture the body takes over from the mind and speaks in its stead. The same holds true of any intense physical stimulus. When Bassanio says "I live upon the rack" (3.2.1389), he does not mean that he is in physical pain.

He means that his life is dominated by his passion (or possibly his lust). We might think of sensual temptation as the diametrical opposite of torture, but it involves a similar attack on the autonomy of consciousness by the manipulation of the body. In Milton's *Comus,* for example, the lady's heroic virtue consists in her refusal to allow her conscience to be violated by Comus's magical seduction. Although his effect on the body is very different from the torturer's, the seducer shares the assumption that the soul can be controlled by physical means. Thus the lady scorns Comus's temptations as "fit to ensnare a brute" (700) and affirms the inviolate integrity of her interior self: "Thou canst not touch the freedom of my mind/ With all thy charms, although this corporal rind/ Thou hast immanacled" (653–655). John Donne conveys the same message in "Love's Exchange." This poem begins by invoking "love" as a malign, deceitful deity, cheating the speaker in a Faustian pact: "Love, any devil else but you/ Would for a given soul give something too" (1–2). This Satanic Eros gives nothing in return for his victim's soul but "Torture." His alleged motive is to "[k]ill and dissect" (40) the speaker, in order to display him as an "example" to "future rebels." However, Cupid's intention that others may "learn by my being cut up and torn" (39) will be frustrated, for no authentic knowledge can be gained by torture. As the speaker informs the blind demon: "this Torture against thine own end is;/ Rack'd carcasses make ill anatomies" (41–2).

The frequency with which the imagery of torture appears in Renaissance English literature, and the capaciousness of its applications, seems to confirm the thesis that torture played a significant role in the fashioning of the early modern subject. According to Nietzsche's *Genealogy of Morals*, the judicial infliction of physical pain provided the conceptual location for such experiences as guilt, will, and responsibility, which are envisaged as residing within the conscience. By addressing itself to an interior entity, which it systematically differentiates from the self's physical expressions, torture forcibly hollows out the space to be filled by an inner, nonapparent subjectivity. It has been persuasively argued that the imperative of fashioning such an internal, conscientious subject provides an explanation for the "witch craze" of the sixteenth and seventeenth centuries. Such arguments take their cue from Foucault's observation that the Enlightenment shift from physical to psychological discipline and punishment assumes a conscious, motivated psyche available for chastisement.

In my opinion, however, it may make more sense to see torture as an attempt to *destroy* the interior subject by forcing it into expression. Perhaps torture does not create the internal subject but rather drags that subject into the light—that is to say, objectifies it. When a substance that is defined by its spiritual nature is forced into physical expression, it loses its definitive characteristic and thus ceases to exist. In this sense, we might even conceive of torture as soul-destroying as opposed to self-fashioning. Shakespeare is certainly careful to specify when, and to what degree, torture affects the psychology of its victims. In *The Winter's Tale* the distinction between physical and psychological torture is woven into the dialogue between Paulina and Leontes. She scorns his power to inflict bodily torture: "What studied torments, tyrant, hast for me?/ What wheels? Racks? Fires? What flaying? Boiling?/ In leads or oils? What old or newer torture/ Must I receive?" (3.2.174–177).

Her mockery springs from her knowledge that she is inflicting far worse psychological torture on Leontes by her relentless reminders of his guilt. In contrast, physical torture presupposes a material subject. Physical torture abolishes mediation between soma and psyche, seeking to access the latter via the former. Indeed, Shakespeare sometimes imagines the torture victim's soul as existing apart from the tortured body. In *Antony and Cleopatra,* Antony suggests that Caesar can take revenge on him by torturing "Hipparchus, my enfranched bondman, whom/ He may at pleasure whip, or hang, or torture,/ As he shall like, to quit me" (3.13.146–149). As described in Nietzsche's *Genealogy*, Antony expresses psychological guilt in terms of an equivalent amount of physical pain. The fact that such pain can be displaced onto a surrogate differentiates the tortured body from internal subjectivity. In contrast, Leontes's torture is exclusively internal, consisting solely in Paulina's interpellation of his guilty conscience.

The drunken "Porter of Hell-gate" in *Macbeth* makes the same distinction between external expression and interior subjectivity when he welcomes Father Garnett to his eternal abode. The Jesuit is described as one who utilized his linguistic dexterity to powerful effect in life, "yet could not equivocate to heaven" (2.3.9), since the contents of the mind cannot be concealed from the Almighty. This returns us to the figure of Proteus who, in addition to symbolizing nature, was also a conventional symbol of verbal equivocation. The *Treatise of Equivocation,* discovered in Garnett's possession, warns against the exploitation of ambiguity by opportunists "more inconstant than *Proteus.*" In order to equivocate, one must distinguish between the inner, hidden self and external, manifest truth. It must be possible for truth to exist prior to and apart from its expression. In rhetoric, the notion of equivocation rests upon the concept of a "mixed proposition." In Aristotle, such a proposition is part verbal and part mental; hence it is possible to use the verbal part without the mental and to speak misleadingly without lying. Sir Edward Coke, prosecutor at the trial of Father Garnett, defined equivocation as: "the pretext of the lawfulness of a mixt proposition to express one part of a man's mind, and retain another."[25]

However, the *Treatise on Equivocation* puts it rather differently. The anonymous author says of a "mixt proposition" that "part of it is expressed, part reserved in the mynde."[26] Here the "mynde" is an element of the self that is not expressed at all, whereas for Coke the expressed proposition is itself an element of the mind, just like the unexpressed. Coke defines equivocation as a distinction between two parts of the mind, while the *Treatise* understands it as a distinction between the mind and verbal expression. The *Treatise* goes on to argue that a "mixt proposition" is not a lie because God sees into the mind and knows the full statement to be true. The "mynde" is pointedly differentiated from "the signes which doe expresse our mynde, partly in speaking and partly in writing," so that an alteration in the mere signs does not affect "the verity of the position."[27] From this perspective, truth and falsehood reside ultimately in the soul; they are not effects of representation: "there is never falshood in the voyce but there is first falshood in the mynde. Wheras verety and falsitye are principally in the vnderstanding, and than secundarylye in the voyce, as in an expressive instrument of that which was false in the mynde."[28]

Verbal signs can mean something different from what was intended by their author. Simply by allowing this natural capacity of signs to take effect, the equivocator can give the interrogator a false message without actually lying in his "mynde." The interrogator's job was to thwart this ruse by forcing the "mynde" into verbal expression. Since the "mynde" is defined precisely as that which is not expressed, however, an attempt to force it into expression is also an attempt to destroy it, to kill it, to alter its proper, interior and spiritual essence. The author of the *Treatise* is perfectly comfortable with the fact that this "mynde" is constructed as a by-product of equivocation: "those wordes which are vttered, if they be taken alone, are most false; therfore that we may cleere them of falshood, we must say of necessitye that they be but part of a proposition, the rest beyng reserved in the mynde."[29] The disjunction between the expressed and the unexpressed self is the very definition of torture for the heroine of *The Rape of Lucrece*: "And that deep torture may be call'd a hell/ When more is felt than one hath power to tell" (1287–1288). The existence of an inexpressible self becomes painful to the degree that one feels pressure to express it, and torture is the bringing to bear of such pressure. In seeking to prevent equivocation, torture abolishes the difference between the verbal and the mental elements in a "mixt proposition." By seeking to reduce the spiritual self to a carnal self, by attempting to impose objectification by force, torture direct attacks the interior human essence or soul.

This same logic is at work in Rowley, Dekker, and Ford's *The Witch of Edmonton*. The eponymous witch Mother Sawyer equivocates in an attempt to save her soul, but the devil (in the form of a dog) interrogates her, using the threat of torture to force her to bargain it away, and thus to destroy it:

Mother Sawyer: Then I am thine; at least so much of me
As I can call mine own—
Dog: Equivocations?
Art mine or no? Speak, or I'll tear—
Mother Sawyer: All thine.[30]

The canine demon knows that the expressed or performed self is not identical with the psyche. He therefore refuses to allow Mother Sawyer to save her soul by equivocation and insists that she eliminate the difference between the performative and the essential self. By doing so, she kills her soul. In this play, which appears to reflect the reality of English witch hunts, torture does not produce an internal, essential subjectivity but rather destroys it.

In conclusion, it appears that a subtle, subterranean debate is being conducted within and beyond Shakespeare's plays, in which Bacon's trope of the scientist as torturer is invoked in various contexts and evaluated in a variety of manners. Perhaps the valence we give to that metaphor finally depends on our attitude toward the autonomous human subject. If, like many postmodern historians of ideas and literary critics, we regard the nonmaterial self as an ideological illusion designed to serve the interests of normative, patriarchal power, we will find it easy to believe that it is, wholly or in part, the product of torture. If, in contrast, we regard the autonomous subject as an ontological reality under attack by ideological mechanisms of objectification and reification, we will find it more plausible that torture is complicit in that subject's destruction.

This is not the place to debate which of these conceptions of subjectivity is most accurate. But it may be an appropriate moment to recall that belief in a nonmaterial, essential subject or "soul" was practically universal in Shakespeare's day, and that there is no reason to doubt he shared it. In that case, we can venture the conclusion, which on the whole seems vindicated by the admittedly ambivalent evidence of the plays, that Shakespeare was troubled by Bacon's instrumental view of nature and sought to criticize it at strategic points in his own work. This does not mean that Shakespeare opposed Bacon; indeed, similar doubts were shared by Bacon himself. It does, however, reveal a worldview at odds with the twenty-first-century consensus (shared by humanists and scientists) that the human self is material and contingent, identical with the body and the signifying systems that mediate our experience. We might do well to reconsider the reasons for Shakespeare's equivocal, vexed, tortuous response to Bacon's infamous image.

NOTES

1. Alexander Smith, "On the Writing of Essays," in *Dreamthorp* (Oxford: Oxford University Press, 1914), 25.
2. Thomas Carlyle, "The Hero as Poet," in *Heroes and Hero-Worship* (London: Chapman and Hall), 27.
3. Gerald Massey, *The Secret Drama of Shakespeare's Sonnets* (Richard Clay, 1888), 380.
4. Francis Bacon, *The Great Instauration*, in *The Works of Francis Bacon*, ed. James Spedding (New York: Garrett Press, 1968).
5. *The Works of Francis Bacon*, ed. James Spedding (New York: Garrett Press, 1968): 6:726, 6:725.
6. In Virgil's *Georgics*, for example, Aristaeus wrestles with Proteus in order to discover the reasons for the illness of his bees.

7. Sir Thomas Browne, *The Garden of Cyrus* (1658).
8. Anonymous, *A Treatise of Equivocation,* ed. David Jardine (London: Longman, Brown, Green, and Longmans, 1851), 54.
9. Cıted in Rudiger Safranski, *Schoepenhauer and the Wild Years of Philosophy*, trans. Edward Osers (Cambridge, Massachusetts: Harvard University Press, 1991), 181.
10. Seminal texts using these approaches include Evelyn Fox Keller, *Reflections on Gender and Science* (New Haven, Connecticut: Yale University Press, 1985); Robert Watson, *Back to Nature: The Green and the Real in the Late Renaissance* (Philadelphia: University of Pennsylvania Press, 2007); and Theodor Adorno and Max Horkheimer, *Dialectic of Enlightenment* (New York: Continuum, 1969). For a more recent, subtle, and nuanced treatment, see Sophie Weeks, "Francis Bacon and the Art-Nature Distinction," *Ambix* 54:2 (July 2007): 101–129. See also William R. Newman, "Alchemical and Baconian Views on the Art/Nature Division," in *Reading the Book of Nature: The Other Side of the Scientific Revolution*, ed. Allen G. Debus and Michael T. Wlaton (Kirksville, Missouri: Sixteenth Century Journal Publishers, 1998): 81–90.
11. Carolyn Merchant, *The Death of Nature* (San Francisco: Harper & Row, 1980), 168.
12. Elizabeth Hanson, *Discovering the Subject in Renaissance England* (Cambridge: Cambridge University Press, 1998), 25. See also Elizabeth Hanson, "Torture and Truth in Renaissance England," *Representations* 34 (Spring 1991): 53–84.
13. The feminist strand of this critique can be traced from Iddo Landau, "Feminist Criticism of Metaphors in Bacon's Philosophy of Science," *Philosophy* 73: 283 (Jan. 1998): 47–61, through Carolyn Merchant, *Re-inventing Eden: The Fate of Nature in Modern Culture* (New York: Routledge, 2004): to Carolyn Merchant, "The Secrets of Nature: The Bacon Debates Revisited," *Journal of the History of Ideas* 69:1 (2008): 147–162.
14. Peter Pesic, "Francis Bacon and the "Torture" of Nature," *Isis* 90:1 (March 1999): 81–94; "Proteus Unbound: Francis Bacon's Successors and the Defence of Experiment," *Studies in Philology* 98:4 (Autumn 2001): 428–456, and "Proteus Rebound: Reconsidering the 'Torture' of Nature," *Isis* 99:2 (June 2008): 304–317.
15. Pesic, 81. Anthony Funari has recently pointed out that Proteus is by no means innocent of antagonism toward his interrogator: "Bacon characterizes Nature as aggressive and resistant to human endeavors of restoration, presenting obscured images of itself to the scientific gaze"—*Francis Bacon and the Seventeenth-Century Intellectual Discourse* (New York: Palgrave, 2011), 116.
16. Cited in Pesic, 92.
17. See David Hawkes, *John Milton: A Hero for Our Time* (New York: Counterpoint Press, 2009) and "Introduction" to John Milton, *Paradise Lost* (New York: Barnes and Noble, 2004).
18. Ayanna Thompson, *Representing Race and Torture on the Early Modern Stage* (New York: Routledge, 2009).
19. Aristotle, *Politics*, trans. Carnes Lord (Chicago: University of Chicago Press, 1984), 1252b1.
20. In addition to the modern sense, however, the word *kind* meant "natural": intriguingly, these lines could suggest that Caliban is being treated unnaturally by being enslaved.
21. Elaine Scarry, *The Body in Pain: The Making and Unmaking of the World* (Oxford: Oxford University Press, 1985), 38.

22. For the background to the current debate, see John Langbein, *Torture and the Law of Proof* (Chicago: University of Chicago Press, 1976); James Heath, *Torture and English Law: An Administrative and Legal History from the Plantagenets to the Stuarts* (Westport, Connecticut: Greenwood Press, 1982); Edward Peters, *Torture* (Philadelphia: University of Pennsylvania Press, 1985). For an application of established theory to the changed circumstances of the third millennium, see Alfred McCoy, *A Question of Torture: CIA Interrogation from the Cold War to the War on Terror* (New York: Metropolitan Books, 2006).
23. Alan Dershowitz, "Want to Torture? Get a Warrant," *San Francisco Chronicle*, January 22, 2002.
24. Jonathan Alter, "Time to Think about Torture," *Newsweek,* November 4, 2001.
25. Cited in Frank L. Huntley, "*Macbeth* and the Background of Jesuitical Equivocation," *PMLA* 79:4 (1964): 390–400; esp. 390.
26. *Treatise*, 9.
27. *Treatise*, 13.
28. *Treatise*, 14.
29. *Treatise*, 14.
30. *The Dramatic Works of Massinger and Ford,* ed. Hartley Coleridge (London: Bradbury and Evans), 191.

2 Plays, Playing, and Make-Believe
Thinking and Feeling in Shakespearean Drama

Ros King

PREAMBLE

When our daughter was about six, we took her to see *Much Ado About Nothing* at the Barbican in London, with Derek Jacobi playing Benedick.[1] Toward the end, following his grumble that he could "find out no rhyme to "lady" but "baby'" he paused in discontent. This pause was matched with silence from the audience; a silence suddenly pierced by a delighted cackle from our child. Jacobi gestured in her direction with a slight shrug before continuing, "an innocent rhyme," which brought the house down (5.2.32).[2]

So what was happening here? The grownups, educated in the conventions of the darkened auditorium (this was a good thirteen years pre-Globe), probably took their cue from the emotion projected by the actor. Our daughter, less well versed in what was expected (she had been frightened by the pirates in *Peter Pan* only eighteen months previously) but probably now thinking of herself as much more lady than baby, saw the funny side of the linguistic collocation. Jacobi allowed the specific reality of the moment to intrude on the fiction, incorporating a pun at our daughter's expense: "some innocent!" We the audience laughed; but at what? The sudden shock of the gap between childish raucousness and polite RSC theatregoing? The latent ludicrousness of both the collocation and Benedick's frustration? We proud parents for our star child in identifying a joke? Annoyance that a young child should be up late and intruding on adult pleasures? Jacobi's skill in turning a near heckle to his advantage? All or none of the above? Or probably all of the above and some, in differing amounts by different individuals, as we all struggled to make sense, repeatedly shifting our focus between the world of the play and the world of the auditorium. The burst of laughter indicated a shared experience, prompted by the words of the not-quite and rather ludicrous rhyme, but not confined by them.

This chapter explores the complex relationships set up between stage and auditorium, word and gesture, and between mind and body, every time we experience a play in performance. I consider some of the philosophical, physiological, historical, and educational practices and values of playing and playfulness; the relationship of play to the development of mindfulness; and the performance of plays as the physical expression of feelings. I conclude with a case study concerning Shakespeare's use of the concept of play in *Much Ado*. My purpose, then, is to demonstrate that the works of Shakespeare—which are so often presented as, on the one hand heritage culture, and on the other as difficult literary texts to be set for examination—demonstrate a deep understanding of the form and function of play in all human activity.

THE PHILOSOPHY OF PLAY

Why do human beings watch plays? Other species make spectacles—puffing out their chests, or raising crests or tail feathers, either to attract a mate or to warn off a rival. All baby mammals play, seemingly as practice for adult behaviors. But we engage in make-believe. We enact other living beings—from historical persons to gods and goddesses, animals to aliens—and enjoy watching others do so. We can also identify with such performances, investing them with our feelings, even when we know their embodiments are inanimate puppets, cartoon characters, or even on occasion abstract squiggles; they are, in Kendall Walton's phrase, "props in games of make-believe" that we play with ourselves.[3] While the personages created, and their style, will vary according to culture and historical period (and there are societies where drama or other forms of art have been suppressed entirely for reasons of religion or politics), the need both to make and to watch representations of life forms seems, at the very least, deeply engrained.

The idea that the gods look down to observe dispassionately the actions of men as a play is attributed to Pythagoras and found repeatedly in classical Greek and Roman literature. According to Cicero, Pythagoras even said that the reason wise men went to the games was in order to watch the behavior of other people and "scan the nature of things."[4] The concept "All the world's a stage and all the men and women merely players" was thus already ancient when Shakespeare penned that line for the melancholy and therefore ostensibly *un*playful traveler Jacques in *As You Like It*. Because he *is* a traveler, of course, and personally uninvested in the people he encounters, the life he observes is like a play unfolding before his eyes. He revels in its incongruities, using them as the motivation for "philosophical" musings—"I met a *fool* in the forest" (2.7.11; my emphasis). Of course this encounter is neither as incongruous nor quite so unproblematically hilarious to us. We know exactly why Touchstone is wandering the forest, a veritable fish out of water, and we might rather be laughing at Jacques's lack of understanding. In the same way, in the next scene but one, after a reminder of the cruel court from which they have all fled, we can enjoy Corin and Touchstone arguing about the appropriateness of different types of courtly or rustic behavior in context: "you kiss your hands; that courtesy would be uncleanly if courtiers were shepherds" (3.1.44).

The *Oxford English Dictionary* gives a staggering 2,414 definitions of the word *play*, illustrated by 13,339 quotations. To give the briefest of tastes of the range involved, these definitions and phrases include "brisk movement" (Old English); "in play," "in occupation or operation," "in combat"; "the play of light, and of weapons"; "freedom or room for action"; "pleasure, recreation, dalliance, sexual foreplay"; "pageantry, games, sport"; "contest"; "drama". Locked into the history and usage of this word, therefore, is an indication that play is a central part of human activity, although the very range of its manifestation is a cause for anxiety because of the many occasions in which playfulness plays or otherwise crosses swords with notions of value and of morality. As Plato's Athenian puts it in a passage in the *Laws*—which encompasses many of the definitions given above—man was

> created as a toy for God. What, then, will be the right way to live? A man should spend his whole life at "play"—sacrificing, singing, dancing—so that he can win the favour of the gods, and protect himself from his enemies and conquer them in battle. He'll achieve those aims if he sings and dances in the way we've outlined; his path, so to speak, has been marked out for him.[5]

The Athenian allows prescribed singing and dancing as a form of religious devotion, presumably because such acts are socially cohesive. But he has severe reservations about unrestricted children's play because it encourages free thinking:

> *Athenian:* I maintain that no one in any state has really grasped that children's games affect legislation. . . . If you control the way children play, and the same children always play the same games under the same rules and in the same conditions, and get pleasure from the same toys, you'll find that the conventions of adult life too are left in peace without alteration. But in fact games are always being changed and constantly modified and new ones invented. . . . Listen: I'll tell you just how big an evil I maintain it is.
>
> *Clenias:* You mean the way the public grumbles at old-fashioned ways of doing things?
>
> *Athenian:* Exactly.[6]

Plato's philosophy is itself somewhat playfully written in the form of a dramatic dialogue. His Athenian may or may not be his idea of what Socrates thought, and his own thought may lie in the gaps between the Athenian and his interlocutors, although the unease is apparent. Writing in response to Plato, Aristotle's "tangled and contentious"[7] treatment of play univocally expresses the moral dilemma:

> If we need both work and leisure, but the latter is preferable to the former and is its end, we must ask ourselves what are the proper activities of leisure. Obviously not play; for that would inevitably be to make play our end in life, which is impossible. Play has its uses but they belong rather to the sphere of work; for he who toils needs rest, and play is a way of resting, while work is inseparable from toil and strain.[8]

Aristotle then goes on to argue for "a form of education which we must provide for our sons, not as being useful or essential but as elevated and worthy of free men . . . because it teaches one to be observant of physical beauty."[9] He means gymnastics, and certain prescribed types of engagement with *mousike*—or those technical "skills presided over by Apollo and the Muses"[10]—always remembering, of course, "that we are not to educate the young with a view to their amusement. Learning brings pain."[11]

Thus, in the period before Descartes triggered such a long-lasting division between body and mind, schoolmasters, the ferrule-bearing tormentors of little boys struggling to learn their Latin, enjoyed the title *ludi magister*, and on occasion wrote the plays, both in Latin and in English, to be performed by their boys.[12] Humanist educators argued for the usefulness of playacting; it trained the memory, gave boys confidence in the important art of public speaking, and also constituted a necessary antidote to and refreshment from toil. Nevertheless, playacting straddles a fault line in moral philosophy and human behavior. On the one hand it is a mechanism for learning about the world and making our societies work. On the other it can be variously coercive, a tool for social constraint (behaving in the "right" way), or subversive, a threat to the existing order, categorized (depending on your point of view) as either immoral and duplicitous or as a mechanism for social change and betterment.

A keyword library search using the word *play* shows that current usage comes under three main groupings: the playing of musical instruments or of cards and games, plays as a literary category, and the role of play in the early years education. Play in the last context demands extended time, dedicated space, and freedom from interference and interruption, although these books are usually at pains to argue that adults need to provide the opportunity and can enrich the process through careful intervention. Over the age of five (in the United Kingdom), playtime is something usually reserved for the breaks between lessons and not generally regarded as itself part of the education process. There is reluctance to allow children at all levels the freedom to explore for themselves that play affords.

In the preface to the third edition of her seminal edited collection *The Excellence of Play*, Janet Moyles writes in some despair:

> It's surprising—and more than a little distressing to me—that . . . we should still be questioning the value of play in children's learning, especially since so much has been happening about play in the education and care of young children in the intervening years.[13]

She states that the emphasis on formal learning, target setting, and assessment higher up the education system is affecting even the under-five curriculum and argues that too few early years practitioners sufficiently understand the scientific evidence for the importance of play in child development. This may be an inadvertent side effect of government policy rather than the policy itself. A U.K. government paper produced in 2009 noted that "a playful approach supports learning because: playful children use and apply their knowledge, skills and understanding in different ways and in different contexts," thus developing flexible thinking and self-motivation.[14] Sadly, these essential life skills are frequently underdeveloped in even the best-qualified undergraduates. Nevertheless, because of the long-standing antithesis with "work," it seems that no government policy is overtly going to embed play into the curriculum.

Johan Huizinga has been one of the few modern writers to explore the qualities of wider human play as opposed simply to that of young children, and in *Homo Ludens* (1944) explores the Renaissance concept of *serio ludere*, or serious play. He points out that seriousness is a weak noun, derived from an adjective, whereas play is the thing itself:

> The significance of "play" . . . is by no means defined or exhausted by calling it "not earnest," or "not serious." Play is a thing by itself. The play-concept as such is of a higher order than is seriousness. For seriousness seeks to exclude play, whereas play can very well include seriousness.[15]

To the amusement of adults, young children are usually deeply serious about their extended periods of free imaginative play even while being aware that it is "just pretend." That seriousness is justified. The ability to appraise those different levels of existence is a concomitant of the development of mind, of social responsibility, and of empathy for others, or what Howard Gardner terms interpersonal and intrapersonal intelligences.[16]

Aristotle's acceptance of play into the realm of aesthetic education was developed by Hans-Georg Gadamer in *Truth and Method* (1960), but primarily in relation to the art object itself:

> When we speak of play in reference to the experience of art, this means neither the orientation nor even the state of mind of the creator or of those enjoying the work of art, nor the freedom of a subjectivity engaged in play, but the mode of being of the work of art itself . . . the work of art has its true being in the fact that it becomes an experience that changes the person who experiences it.[17]

But this quest for the methodological "truth" of art seems to allow insufficient autonomy to receivers of artwork to make of it what we will, and to use it as a "prop" for our own private imaginings. The consciousness that what we are watching is "only pretend" allows a multiplicity of "truths."

Gadamer goes on to argue that "the primacy of play over the consciousness of the player" is such that "the structure of play absorbs the player into itself, and thus frees him from the burden of taking the initiative, which constitutes the actual strain of existence." Again, I remain unconvinced. From personal experience, the play of four-year-olds involves constant negotiation and initiative-taking: "You be the monster, granny, and I be the knight," which is clearly not just about getting lost in a game of dungeons and dragons but about exerting autonomy in family relationships.

THE PERFORMANCE OF SCRIPTED DRAMA

The performance of a written play text consists of at least three separate but interlocking and mutually dependent plays or games. The text itself is a playground for the imaginative free flow of its writer, in negotiation with any co-writer, and with the commissioning company of actors or producers. It is a game of sources, influences, practicalities, and words. The production is a game devised in response to that text by its actors and those involved in the staging; in the case of productions of classic plays, the artists may also be playing consciously and unconsciously with previous productions and performances, whether they are trying to do something "new," or paying homage to the past.[18] Those two plays together create a third play: the unique, unseen, and unknown play in the head of each spectator, which may not be just concerned with what is being performed but may include reflection on that individual spectator's current concerns, or previous and immediate life experiences, combined with conscious awareness of the performance as distinct from the story.

Hamlet, for example, muses on the player's performance of the Hecuba speech and wonders why his own behavior fails to match up. The player, with "tears . . . distraction . . . a broken voice, and his whole function suiting/ With forms to his conceit," has forced his very soul to match his "dream of passion," making Hamlet berate himself for doing "nothing": "What would he do,/ Had he the motive and the cue for passion that *I* have?" (2.2.545–555, my emphasis). Claudius is similarly aware of different levels of "truth" and "pretend" at the performance of *The Mousetrap*. He has no emotional response at all to the dumb show. Murder stories are common enough in entertainment, after all, and this is an old tale, performed in a stylized way. It touches him not and we do not have to suppose (as generations of critics and performers have done) that

and their gestural manifestations can be altered according to societal norms. It is this distinction that I am observing here.

Strong feelings for and about other individuals are a necessary concomitant of social behavior and for the rearing of the young. The human infant is usually born into a world where its facial expressions are imitated as a form of loving, socializing interaction by its carers, and where it is later encouraged by its teachers to practice high-level skills through imitation. We are both equipped and conditioned to derive meaning from gesture. But again we have only recently become aware of the extent to which such meaning-making is preconscious. So-called mirror neurons that are activated in reaction to observed movement in other individuals were first identified in the brains of macaque monkeys in 1996.[26] While the evident difference between monkeys and humans has led some to question the precise function of these neurons as suggested by their earliest discoverers, there is now increasing evidence that a "mirror neuron system" in humans may be a "by-product of associative learning and contribute to a range of social cognitive functions."[27] But whatever the process whereby mirror neurons are developed, they seem to be related to motor neurons and to cause the firing of those specific motor neurons that would be activated were we ourselves to be making similar movements, whether or not we are doing so.

Thus, when we observe a nonhuman social animal such as an elephant or a horse whose offspring has died, listlessly hanging her head or fruitlessly attempting to nudge the dead infant back to life, the physical gestures are recognisable to us as expressions of loss, and even of what *we* experience as love. They are therefore capable of arousing in us intense feelings of pain. I now even feel differently about crocodiles after watching one enormous and ferocious creature gently rocking her eggs through the water between her teeth in order to help free the struggling hatchlings, using movements that in a human would be characterized as parental bonding. The stern injunction not to anthropomorphize in such situations is missing the point. No matter what the animal itself is conscious of feeling (and the crocodile would no doubt simply see *me* as dinner), we derive meaning for ourselves from observing the movement of others in context. It seems that the embodiment of the brain involves incessant instantaneous feedback loops: involuntary response to an environmental stimulus creates endocrinal change and movement; awareness of this physicality constitutes feeling; feelings can be mentally compared and mediated; observation of the movement of others prompts firing of the motor neurons in ourselves; this may result in unconscious movement and/or in an emotional endocrinal response; and so endlessly on.

Thus, when we watch actors performing in front of us, the play is also going on in our heads, not only because it awakens conscious memories of similar events in our own lives but also because it is operating on us on an unconscious neural level. The question is complicated because human physical expression and gesture are often consciously developed, or controlled and self-censored, according to social pressures or simply through exposure to different ways of doing things. Film and sound recordings made over the last 100 years show just how much day-to-day intonation and gesture have changed over a relatively short period. Performances in plays and films that are more than a few years old can seem dated, even stilted and unnatural, even though the common refrain in acting manuals over 200 years (and even going back to Hamlet's advice to the players) is to "hold the mirror up to nature," or rather to "show . . . the very age and body of the time his form and pressure" (3.2.20; 22–25). We may need historical iconographical research to read the gestures recorded in Renaissance paintings,[28] but it should not necessarily be assumed that changing forms of expression imply changes in emotion or indeed feelings. Who would have thought that the stiff upper-lipped English of just 60 years ago familiar from all those 1940s and 50s war films can now be found performing grief so extravagantly with bouquets, candles, and teddy bears piled at the scene of a death? Performance is catching. Or rather, perhaps, ways of performing grief, which the stiff upper lip was designed to suppress as working class and potentially subversive—but which can be seen in early silent movies, and discerned from illustrations from the sensationalist, melodramatic popular theatre of the nineteenth century—have now been given the oxygen of Internet ubiquity.

WORDS, SPEECH, AND GESTURE

With their peculiar concentration of images connected with movement, Shakespeare's plays work on the interface between the conscious and the unconscious mind of both actor and audience:

> And pity like a naked new-born babe,
> Striding the blast, or heaven's cherubim horsed
> Upon the sightless couriers of the air,
> Shall blow the horrid deed in every eye,
> That tears shall drown the wind. (*Macbeth*, 1.7.22–25)

These apocalyptic images are not just there for decoration or literary effect. The active visualization of them by the actor will help him arouse feelings (here, of both pity *and* terror) that will be expressed in his voice. The speech runs through a gamut of different feelings in succession but achieves its final tumultuous effect through an extreme intermodality of sight and sound that confuses the two. Vision of the horrid deed fills the eye, causing an automatic physical reflex and an expression of felt emotion so extreme that no other sensation is possible; just as the eyes are blinded by being filled by the deed and the tears that result, so the sound and touch of the wind that brings that vision are drowned out.

Were the actor to be using a memory system[29] to help him with that punishing routine of having to remember and perform six different plays per week, every week, he might perhaps place mental images of the babe and the horsed cherubim on different architectural features of his memory theatre, mirrored in the auditorium before him. In the act of recall, he would be moving through that imaginary building, and perhaps also in actuality across the stage—somatic or muscle memory aiding his recall of the words. He might also be employing the gestures advocated in the books of classical rhetoric, pointing out those imaginary creatures to his audience, and then, on the cue "in every eye," perhaps eyeballing individuals in the audience in both vertical and horizontal planes. The audience in turn, consciously picturing those moving images and witnessing the moving actor, will be unconsciously mirroring the actor's movements on a neural level, and involuntarily even on a muscular level.

And yet one's conscious physical awareness of oneself as a spectator in a theatre audience prevents one from being overwhelmed by this unconscious physiological, emotional reaction to a fictional situation. Thus, in a poem apostrophizing greatness, John Davies of Hereford, writing tutor to the nobility, compares his own middle or "mean" station in life to that of a spectator in the theatre, leaning on a pillar amongt a crowd or "press" of people in the middle gallery. His overt intention is to express the advantages attached to *not* being great. Inadvertently, he gives us a brilliant description of the way we experience theatre:

> . . . like a looker on a tragedy
> Within the middle room, among the mean,
> I see the fall of state and majesty
> While 'mongst the press t'a pillar sure I lean:
> So see I others' sorrows with delight,
> Though others' sorrows do but make me sad:
> But plagues to see, which on our selves might light,
> Free from their fall, makes nature, grieving, glad.[30]

Whatever it was he saw, however it was performed, and despite his very firm early modern notions of social status and his culturally specific understanding of a pillar as a symbol of fortitude, the physical and emotional but also feeling and intellectual or reflective process he describes are certainly recognizable to me. It is the material pressure of the pillar, however emblematic it might also be, that is the telling touch; it draws attention to his physical consciousness. While being moved by and emotionally engaged with fictional others he remains, through this reminder of his physical presence in the theatre, acutely, and feelingly, aware of himself and his personal situation. It is a telling paradox, and shows how the very physicality of the theatrical experience enables the spectator to be mentally fully present in more than one realm simultaneously, able to empathize with other people from other times and places (even from no time or place) without being subsumed by that identity.

The awareness that a game of make-believe is just that—make-believe—is not a cause for regret, but the source of the delight, whether the story itself is happy or sad, frightening or funny. This then has little to do with the famous definition of theatre by the early-nineteenth-century poet Coleridge as a "suspension of disbelief." That may have worked as an explanation for the illusionist theatre of his day, where audiences were expected to marvel at the "realism" of painted sets, but it does not adequately explain the experience of watching plays performed in the open-air theatres of Shakespeare's London, where the fun, and the cognitive value, lies in being reminded constantly that one is in a theatre, engaged in make-believe. Davies pits his own thoughts and feelings against those he sees enacted on the stage. He is probably correct; we enjoy watching misfortune when it reinforces our own sense of security. We feel empathy for these "people" playing out their lives in front of us, even though we know they are figments of imagination; we watch the actors present the characters, yet the act of watching encourages us to think about what we would do in a similar situation; we are sad while knowing that we have no reason to be sad about this particular story, since it is a fiction, even if based on true events; we might sympathize with the characters while simultaneously admiring the actor's skill. We will be engaged by the story; we may even be implicated in it, when villains like Richard, Duke of Gloucester in *Richard III* or Iago in *Othello* take us into their confidence and spark our enjoyment in their plots and murders. At the same time, that constant physical consciousness of ourselves (the pressure of the pillar) reminds us that our lives are separate from what we see depicted and enables us to maintain a healthy, skeptical distance. This ambivalence is important and is what distinguishes the theatrical experience on the stage from theatrics in everyday life, where we, as ourselves, might be swept along, for example, by a demagogue's oratory, becoming pawns in that individual's personal ambitions for real.

MUCH ADO ABOUT NOTHING

The basics of life—being born, avoiding hunger, fleeing danger, reproducing, dying—are not play. And they are too fundamental to be graced with the epithet *serious*; they just are; to paraphrase *Hamlet's* Gertrude, all that lives must struggle to live and then die. Everything else, everything that involves reflection, creativity, choice, skill—everything that makes life intellectually and feelingly interesting—is play. All Shakespeare's plays contain a high frequency of references to playing and acting, but *Much Ado About Nothing*, as signaled by its title's claim to insignificance, seems to have recognized this aspect of evolved life more than most. It is a series of games and playacts from start to finish—many of them deadly if not exactly serious. The play begins with Leonato, Governor of Messina, talking to a messenger who has brought him news that Don Pedro and his troops, who have recently won a notable victory in the wars, are about to descend on his hospitality en masse. Quite what these wars have been about, who they have been against, or what this victory has settled is never established. It is surely stretching the evidence too far to stipulate, as the notes to the RSC edition do, that the vanquished party is Don John.[31] At no point does the dialogue confirm that the idea that he is now "reconciled to the prince [his] brother" is the result of being defeated in the war. Conrad later tells John, "You have late stood out against your brother, and he hath ta'en you newly into his grace" (1.3.18–19), but this "standing out" does not need to be equated with opposing him in battle, and their entry together in the first scene announces to an audience that they had been on the same side.

The fact that the cause and nature of the war is undetermined is important. History shows that the game of war is an uncertain and ineffective way of settling disputes in the long term. When diplomacy resorts to other means, the outcome is usually unsettled, fragile, and in need of a lot more diplomacy. The language in which this particular war is described is therefore nothing if not a game:

LEONATUS: How many gentlemen have you lost in this action?
MESSENGER: But few of any sort, and none of name.
LEONATUS: A victory is twice itself when the achiever brings home full numbers. (1.1.4–8)

The winning participants in this event have been actors in more senses than one. Young Claudio has acted beyond his years, even beyond his species, having "borne himself beyond the promise of his age, doing in the figure of a lamb, the feats of a lion." The messenger's courtliness in cataloguing his achievements and claiming they are beyond expression inadvertently suggests they are beyond not just expectation but also reality: "he hath, indeed, better bettered expectation than you must expect of me to tell you how." Just what does that signify, exactly? News of the young man's exploits have also had an unfortunate emotional (in the sense of involuntary, physical) effect on those who love him most and joy in his success, to the extent that endocrinal and other secretions are mixed up beyond reason, "even so much that joy could not show itself modest enough without a badge of bitterness."

It is easy to overlook the importance of difficult speeches by minor characters in opening scenes, but Shakespearean characters introduced in their absence with so many ellipses and double negatives invariably turn out to be problematic; compare similar descriptions of Posthumus Leonatus in *Cymbeline*, and the boyhood relationship between Leontes and Polixenes in *The Winter's Tale*. Leonatus's platitudinous reply to the messenger's excess ("How much better is it to weep at joy than to joy at weeping!") is too much for Beatrice—or at least that's how I, were I playing the part, would summon the energy to deliver the outrageous yet simultaneously anxious "I pray you, is Signior Mountanto returned from the wars or no?" Her intervention shifts the scene into a different mode, but her very different take on martial exploits reinforces the gaming aspect I have already noted:

> He set up his bills here in Messina and challenged Cupid at the flight; and my uncle's fool, reading the challenge, subscribed for Cupid, and challenged him at the bird-bolt. I pray you, how many hath he killed and eaten in these wars? But how many hath he killed? For, indeed, I promised to eat all of his killing. (1.1.33–37)

Countering the claim to chivalry of an actual war, she recalls a mock chivalric event in which Benedick was involved, where words and writing were more significant than deeds—"bills," "challenged," "reading," "subscribed'—and were more easily categorized as folly. This does more than merely prepare for the "merry war" and the "skirmish of wit" between Beatrice and Benedick (1.1. 32–37, 52, 54); it suggests that everything of value in this society is "indeed" a form of play, although not everyone is sufficiently clear sighted to recognize it as such. Precariously, and insecurely, they are in a state that anticipates Clausewitz's famous phrase but in reverse: here wit becomes the continuation of war by other means.

WITHDRAWAL FROM PLAY

Just as in *The Merchant of Venice*, the "man who has no music in himself/ Nor is not moved with concord of sweet sounds,/ Is fit for treasons, stratagems, and spoils" (5.1.83–85), so, in *Much Ado*, those who resist playfulness, or who take themselves too seriously, are bound to be involved in, or duped by, plots and deceit. If the importance of play in young children is that it develops social skills and empathy, so those who hold aloof from playfulness, as Don John does, are not just "spoil-sports," to use Gadamer's term,[32] but actively antisocial. The rules of the game dictate that those who are designated as players should take part and not hold back. Don John's first line of dialogue indicates his unwillingness to join in: "I thank you; I am not of many words, but I thank you" (1.1.135). His silences and his refusal to engage in the give and take of social ritual is a recurring theme throughout:

> I cannot hide what I am; I must be sad when I have cause, and smile at no man's jests; eat when I have stomach, and wait for no man's leisure; sleep when I am drowsy, and tend on no man's business; laugh when I am merry and claw no man in his humour. (1.3.11–14)

Erroneously (though fortuitously), he is on one occasion designated "dumbe John" rather than Don John in a Folio stage direction (2.1.71.0).

More usually, plays and games attract audiences—those who, as John Davies perceptively observes above, may be emotionally involved in the action while simultaneously separate from it. Such dual engagement will be shown repeatedly in *Much Ado* in different forms: as when the lords and then the ladies become both participants and witnesses to the plot to bring Beatrice and Benedick together, or when they are unwitting witnesses to a staged event (as when they see Borachio make love to "Hero" at the window). But we may also have here a solution to a textual crux. Infamously, the play contains a silent character, Innogen, wife to Leonatus, who makes two appearances in Folio stage directions at the beginning of the play, before the games become deadly in their seriousness. Silent characters are almost impossible to read. One forgets they are there. On stage, by contrast, they can become a focus of attention, drawing the audience's eyes and ears to what is happening. And in *Much Ado*, which brings into play such a range of wordplay, playacts, and plots, in which so many of the characters become by turns and sometimes simultaneously both actor and audience, the "argument of [their] own scorn" (2.3.9), Innogen's presence as silent witness at the beginning and her absence at the end may be neither an accident nor an oversight. At the start of the play, she is our surrogate, reminding us that we too are watchers. But by the end, the ethical issues that the play has raised are too great to remain encased on the stage. The marriage trick that is arranged to bring the play to a close inevitably raises more questions than it answers—for some audience members. There is no longer a single point of view.

Much Ado is currently much set for A-level examination in the United Kingdom, where it is a popular choice among students. There is nevertheless a gender divide in its reception. Teachers report that boys in all-boys schools love it. It seems the trick solves the problem for them. Girls, however, are more likely to ask "how could she take him back?" It may be objected that that is a "modern" response, although modern experience in fact shows that many women do indeed take back abusive partners: "I love him, he will change" is a beguiling prospect. But the play itself is careful to problematize its supposedly happy ending, presenting its resolutions as a series of rather bleak choices. Benedick still thinks of marriage to Beatrice as something that will "bind me or undo me—one of them" (5.3.20) and Claudio remarks on his "February face,/ So full of frost, of storms, and cloudiness" (5.3.41–42). The outcome has set up a ritual in which the characters feel constrained to act according to the rules: "Which is the lady I must seize upon?" asks Claudio obediently when the ladies enter masked, while he and Benedick are still threatening, albeit playfully, to force a change in the behavior of the other by fighting.

The saving grace is that in conversation Beatrice and Benedick become aware of the play acting that brought them together, although the rhetorical patterning of this exchange momentarily threatens to throw them back on their erstwhile sparring:

Benedick: They swore that you were almost sick for me
Beatrice: They swore that you were well-nigh dead for me.
Benedick: 'Tis no such matter. Then you do not love me?
Beatrice: No, truly, but in friendly recompense.

They seem already almost to have forgotten the shattering, naked truth of their raw emotional encounter immediately after Claudio's rejection of Hero. Instead, their secret love sonnets, each to the other, are produced: "A miracle! here's our own hands against our hearts." (5.4.80–92). As is often the case in Shakespeare's later plays, this play ends with a realization that words are sometimes no match for feelings:

Benedick: I'll tell thee what, Prince: a college of wit-crackers cannot flout me out of my humour. Dost though think I care for a satire or an epigram? No. If a man will be beaten with brains, 'a shall wear nothing handsome about him. In brief, since I do purpose to marry, I will think nothing to any purpose that the world can say against it; and therefore never flout at me for what I have said against it; for man is a giddy thing, and this is my conclusion. (5.4. 99–105)

Yet, at the same time, man is always and inevitably a playful, "giddy thing"; there is nothing for it, therefore, but to "play music. . . . Strike up, pipers. *FINIS*."

CONCLUSION

It is common to describe the engagement of audiences and readers in a work of fiction as emotional involvement with the characters.[33] Who has not fallen in love with Hamlet, or at the very least with the actor playing Hamlet? But although this may happen, it is not why the play exerts perennial fascination. If the performance is working as it should, the emotions embodied on the stage will be reflected in our own bodies, but the work of the pillar, or other physical sensation that makes us aware of ourselves in space, is to ground us in our particular present. We are simultaneously both Hamlet (a Renaissance version of a twelfth-century prince) and ourselves. When we add our neuronal and cognitive experience of every other character with their conflicting moral standpoints or approaches to life, call to mind graphic, emotion-generating images such as being "bounded in a nutshell" and yet "king of infinite space" (2.3.253) and become aware that there may be more than one given reason for a significant event, there is ample scope for the imagination and for a reflection that goes beyond simple identification. Our role as interpreters, teachers and performers, is to open up these possibilities latent in the text, enabling our audiences to see and understand its structure as amply and accurately as possible, thereby ensuring that they too have the fullest opportunity to play.

NOTES

1. RSC, dir. Terry Hands, Stratford (1982) and London, Barbican (1983).
2. References to Shakespeare are from *The Complete Works*, ed. Peter Alexander (London and Glasgow: Collins, 1951).
3. Kendall L. Walton, *Mimesis as Make-Believe: On the Foundations of the Representational Arts* (Cambridge, Massachusetts: Harvard University Press, 1990), 51.
4. Cicero *Tusculan Disputations*, trans. J. E. King (London: Heinemann, 1927), v.3.9; cf. "Pythagoras said that this world was like a stage/ Whereon many play their parts," *Damon and Pythias*, 7.71–72, in *The Works of Richard Edwards*, ed. Ros King (Manchester: Manchester University Press, 2001).
5. Plato, *The Laws*, trans. Trevor J. Saunders (Harmondsworth: Penguin, 1970), vii 803c–e.
6. Plato, 797b.
7. Aristotle, *The Politics*, trans. T.A. Sinclair, revised by Trevor J. Saunders (Harmondsworth: Penguin, 1962; revised edition, 1981), 120.
8. Aristotle, 1337b33.
9. Aristotle, 1338a30–1338b2.
10. Johan Huizinga, *Homo Ludens*, trans. George Steiner (London: Temple Smith, 1970; first published in German in Switzerland, 1944), 183.
11. Aristotle, 1339a26.
12. Their teachers may also, some of them, have enjoyed the whipping rather too much; Nicholas Udall was not the only schoolmaster to be sacked for

his abuse of the children in his care. William Hawkins, author of *Apollo Shroving*, was sued by the parents of one of his charges; Milton's friend Alexander Gill, master of St Paul's school and one of Hawkins's circle, was sacked for excessive cruelty. See Ros King "Disgust and Delight," in *Community Making in Early Stuart Theatre*, ed. Roger Sell et al. (Aldershott: Ashgate, 2014).

13. Janet Moyles, ed., *The Excellence of Play* (Milton Keynes: Open University Press, 2010), 1–2.
14. Cited in Moyles, 14. See also John Harland, Kay Kinder, Pipper Lord, Alison Scott, Ian Schagen, Jo Haynes, with Linda Cusworth, Richard White, and Riana Paola, *Arts Education in Secondary Schools: Effects and Effectiveness* (Slough, Berkshire: National Foundation for Educational Research, 2000).
15. Huizinga, 65.
16. Howard Gardner, *Frames of Mind: The Theory of Multiple Intelligences* (New York: Basic Books, 1983): and *Art, Mind, and Brain: A Cognitive Approach to* Creativity (New York: Basic Books, 1982).
17. Hans-Georg Gadamer, *Truth and Method*, 2nd ed., revised, trans. Joel Weinsheimer and Donald G. Marshall (London: Continuum, 2012), 102–103, 105.
18. The RSC wardrobe contains a shapeless long grey cardigan once worn by Peggy Ashcroft. It is a costume item treasured, almost as a confirmation of their craft, by those actors who still get to wear it. Sarah Bernhardt's performance of *Hamlet* (Adelphi Theatre, June 1899) was castigated on both sides of the Atlantic for her attempt to introduce new business into the role.
19. Thomas Dixon, *From Passions to Emotions: The Creation of a Secular Psychological Category* (Cambridge: Cambridge University Press, 2003); see also Jonathan H. Turner and Jan E. Stets, *The Sociology of Emotions* (Cambridge: Cambridge University Press, 2005); James M. Wilce, *Language and Emotion* (Cambridge: Cambridge University Press, 2009).
20. Flavius Josephus, *The Famous and Memorable Works of Josephus,* trans. Thomas Lodge (London: Bishop, Waterson, Short and Adams, 1602), 388.
21. Thomas Wright, *The Passions of the Mind in General* (London: Valentine Sims, 1604), second edition, 180–181.
22. The suggestion was first made by Charles Darwin in *The Expression of the Emotions in Man and Animals* (London: John Murray, 1872), 366. More recent corroboration was found by Fritz Strack, Leonard L. Martin, and Sabine Stepper, "Inhibiting and Facilitating Conditions of the Human Smile: A Nonobtrusive Test of the Facial Feedback Hypothesis," *Journal of Personality and Social Psychology* 54.5 (1988): 768–777.
23. Daniel M. Gross, *The Secret History of Emotion: from Aristotle's Rhetoric to Modern Brain Science* (Chicago: University of Chicago Press, 2006).
24. Antonio Damasio, *The Feeling of What Happens* (London: Vintage, 2000).
25. The psychologist William James, brother to the novelist Henry, was the first person to describe the preconscious operation of flight and fight emotions in "What Is an Emotion?" *Mind* 9.34 (1884): 188–205.
26. Giacomo Rizzolatti, Luciano Fadiga, Vittorio Gallese, and Leonardo Fogassi, "Premotor Cortex and the Recognition of Motor Actions," *Cognitive Brain Research* 3 (1996): 131–141; Giacomo Rizzolatti, Maddalena Fabbri-Destro, and Luigi Cattaneo, "Mirror Neurons and their Clinical Relevance," *Nature Clinical Practice Neurology* 5 (2009): 24–34.
27. Cecilia Heyes, "Where Do Mirror Neurons Come From?" *Neuroscience and Biobehavioural Reviews* 34.4 (2010): 575–583.

28. Michael Baxandall, *Painting and Experience in Fifteenth-Century Italy*, 2nd ed. (Oxford: Oxford University Press, 1988). John Bulwer's *Chirologia* (London: Thomas Harper, 1644) is a famous early attempt by a physician to identify specific hand gestures as having fixed meaning.
29. Frances Yates, *The Art of Memory* (London: Routledge, 1966); Mary Carruthers, *The Book of Memory* (Cambridge: Cambridge University Press, 1990).
30. John Davies, *Wittes Pilgrimage* (1605), T4r, modernized spelling.
31. William Shakespeare, *Complete Works*, ed. Jonathan Bate and Eric Rasmussen (London: Macmillan, RSC Shakespeare, 2007).
32. Gadamer, 103.
33. Jenefer Robinson, *Deeper Than Reason: Emotion and Its Role in Literature, Music, and Art* (Oxford: Oxford University Press, 2005), 113–117.

Figure 2 "Colericus: 2," by Virgilius Solis, the Elder, (1514–1562). Image A021002, History of Medicine Collection, Courtesy U.S. National Library of Medicine.

3 Warmth and Affection in *1 Henry IV*

Why No One Likes Prince Hal[1]

Emma Firestone

A teacher shows his high school English class a production video of *I Henry IV*. He then asks a student—call her Ella—for any comments she might have on the character of Hal. "I just don't really *like* him," Ella mock-whines. A few giggles peal out around the room and the teacher grimaces and quickly moves on to somebody else. What is wrong with Ella's remark? In brief, it contradicts practically everything that the play has to say about its central role: namely, that wastrel-prince Hal is, indeed must be, the instant darling of audiences everywhere except in France. Hal's innate popular appeal derives from a character type that is equal parts generous rake—whose casual "vices" actually reassure audiences that he deserves a good wife, handsome fortune, and noble future, as youthful profligates of seventeenth-century/ classic Hollywood comedy invariably do—and King David, whose audience guards special knowledge of the surprise in store for his fellow characters. The folk history that Shakespeare draws on for the second tetralogy[2] dictates that *1 Henry IV* move toward a climax in the revelation of Hal as the man who will be the heroic Henry V: just the climax that Shakespeare provides, in fact, in Hal's reformation at the Battle of Shrewsbury. Reporting on Hal's gracious, convincingly princely challenge to Hotspur, Vernon says in frank amazement:

> let me tell the world,
> If he outlive the envy of this day,
> England did never owe so sweet a hope
> So much misconstrued in his wantonness. (5.2.65–68).

Theatre audiences, though, would have been as prepared to be surprised by the emergence of the valorous "new" Hal as filmgoers are when Rocky Balboa really does go the distance against Apollo Creed. There is great pleasure to be gained from the easy gravitation of our sympathies toward figures whose virtues are evident to us alone; this is one of the most easily achieved pleasures of narrative drama, the mainspring of countless fairytales and "rags to riches" biographies, not to mention of *Henry V*'s scenes before

Agincourt. In claiming to be unmoved by Hal's winning trajectory, Ella sounds and probably believes herself to be deliberately provocative.

The more serious problem with her response, though, is that it is frivolous. Admittedly the object of her critique is somewhat unclear: either she means that she finds Hal unlikeable as a man or she thinks that the actor playing the prince failed to please in his role. Either way, Ella engages with features of the drama that the Anglophone critical tradition has long insisted are insignificant. Writing of English playgoing culture before the time of Dryden, Samuel Johnson commented: "The structure of dramatick poems was not then generally understood. Audiences applauded by instinct, and poets perhaps often pleased by chance."[3] Ella responds by instinct, and *instinct* is a bad word. The Ibsen scholar William Archer's elegant treatise "The Ethics of Theatre Criticism" (1886), which frankly discourages "sympathy and antipathy" as bases for critical judgment and situates the play itself—not the performers' contributions—as the rightful object of critical attention, articulates standards of excellence in criticism that persist today.[4] Thus delimited, theatre criticism comes inevitably to resemble literary analysis, with all the characteristic ambitions of that genre. Samuel Cullen Carpenter, one of the first important U.S. theatre critics, impressively expressed these ambitions in the premiere issue of his theatre journal *The Mirror of Taste, and Dramatic Censure*:

> If critical judgment were a matter of physical force or numerical calculation, then indeed the roar of the multitude would be as conclusive in reason, as it too often is in practical effect; *but criticism is a matter of intellectual estimate*; and many acquirements go to the composition of a well-qualified dramatic critic, *to any one of which, but a small number of the auditors of a play can, in the nature of things, have the smallest pretensions*. . . . All the genuine critic will attempt to do, is to hold out those lights, with which his own study, experience, and observation have supplied him, in order to enable the public to discern more clearly what in the play or the actor is worthy of censure or applause—of rejection or adoption.[5]

He moves on to list scholarly requisites for critical practice. The critic should be well enough "versed in philological literature" to be capable of forming "a just estimation" of the playwright's "beauty in style." He should know whether the performance carries sufficient "moral institution." And he should remain cognizant that he "sits in judgment for the security of the public taste."[6]

Carpenter's attitudes, however admirable, reflect a perception of critics as possessed of near-superhuman erudition and discernment and capable of insights utterly inaccessible to the "multitude"—though of course more just and edifying than anything in the purview of this ignoble demographic. This perception, certainly as held by critics themselves, has not altered much today. The American Theater Critics' Association has opened its annual conferences with a reading of director-critic Harold Clurman's "The Complete Critic's Qualifications," which feature, for example, #2, "He should know the history of the theatre from its origins to the present"; #6, "He should know the history of his country and world history: the social thinking of past and present"; and #7, "He should have something like a philosophy, an attitude toward life."[7] The self-regard evident in these maxims, though not necessarily the expansive humanism, persists with a strong defensive edge in article titles such as "A Critic's Case for Critics Who are Actually Critical." The author of the latter has this to say about criticism: "It means making fine distinctions. It means talking about ideas, aesthetics and morality as if these things matter (and they do). It's at base an act of love. Our critical faculties are what make us human."[8] He does not use the phrase "roaring multitudes," but those condescending parentheticals make his attitude clear enough.

Stephen Booth has devoted some space in his oeuvre to skewering this critical tradition for its endemic vagueness and insistence on the dignity of literature (and works in other media that command the critical attention previously accorded only to literature, like *The Wire* or the films of David Lynch) over that of other arts. Booth has characterized his own critical method as "a public demonstration of opposition to the understandable but pernicious assumption that literary criticism should . . . mirror the large-spiritedness and disdain for sweat and petty detail that characterize the works we admire and study."[9] Implicitly in his editorial practice and explicitly elsewhere, Booth insists that literary art, irrespective of its great and irrefutable value, is frivolous:

> We [scholars] are uncomfortable with whispers in our consciousnesses that acknowledge that the primary attraction of *King Lear* is the same in kind as *I Love Lucy*'s. Unless I am as entirely wrong as it would be comfortable to think, any distress you may feel when I lump together *King Lear* and *I Love Lucy* is distress over a threat to the dignity of *Lear* or to the dignity of our enterprise as students of it.[10]

What Ella's assessment of Hal does, of course, is to lump *Lear* in with *Lucy*, threatening the dignity that theatre critics have fought for centuries to get accorded to their responses and profession. Still, those giggles from her classmates, like Booth's "whispers in our consciousness," point to the conclusion that is this chapter's central concern. Prince Hal *is* unlikeable. He is unlikeable *as a character*—as a construct of the dramatic text. The actor who portrayed him was not inept in failing to engage her affections. On the contrary, he was correct, even generous, in resisting the temptation to invest Hal with qualities that *1 Henry IV* declares him to possess but never provides occasion for actors to display. Ella's intuition of Hal's unpleasing nature might well have been based on "antipathy" rather than "fine distinctions," "instinct" rather than "intellect." But this chapter, which brings traditional tools of literary analysis into dialogue with insights from the field of social cognition, aims to validate her intuition and explicate the phenomena in human cognition to which it attests. It is intended as a demonstration of the significance—though not, it is worth noting at the outset, the primacy—of automatic and embodied cognitive processes in apprehending dramatic verse.

Additionally and largely implicitly, the chapter advances two arguments for integrating the perspectives of cognitive sciences into studies of Shakespeare's art. First, it is entirely reasonable to regard critical reactions to characters like Hal as bodies of evidence of social psychology in action, four hundred years in development. One could imagine such a record as potentially of use to scientists working on the cognitive processes driving interpersonal understanding and judgment—for reasons this chapter endeavors to explain. Furthermore, I argue that cognitive studies offer compelling evidence that apprehending and judging theatrical characters operates via the same mechanisms as serve for apprehending and judging humans in live, social settings. This approach may provide a measure of insight into why Shakespeare's characters—linguistic objects, comprising the kind of comprehensive semantic patterning that is distinctly Shakespearean in its density and omnipresence—present themselves to readers and audiences with the conviction and credibility of real human beings.

PHYSICAL CONCEPTIONS

If dramatic critics at large have ample grounds to reject Ella's take on Prince Hal, so too do literary scholars. Specifically, in employing the language of social perception and judgment, Ella indulges in the energetically discredited critical habit of treating fictional beings as if they were cogitating agents.[11] The most familiar form of this transgression involves placing characters in the subject position: "Bottom instinctively acquires a most learned taste, and grows fastidious in the choice of dried peas and bottled hay."[12] But to speak of characters in terms of the affection, attraction, and the trust they earn from audience members commits essentially the same conceptual error, since these terms imply that the character-audience relationship is somehow an interactive one and, by extension, that the character possesses agency and subjectivity. Since the behavior, decisions, and destiny of dramatic characters are predetermined by the playwright, though, this obviously makes no sense. It is a critical commonplace that audiences "respond" to characters, but these responses cannot logically be construed as *social* responses; social behavior cannot be limited to one party only. That is the (largely implicit) conventional understanding of why it is incorrect to accord the same kind of evaluative language and judgment to human beings and fictional simulations of them. Sustaining this understanding is, of course, our investment in the idea that most or all significant human behavior and judgment is consciously, rationally motivated.[13] This is a cornerstone of Cartesian and Lockeian philosophies of mind and persistent in fields (humanities prime among them) concerned with the content and productivity of human intellection. Intellect informs us that to perceive a character as a thinking subject is erroneous, just as it is erroneous to try to stop Othello on his murderous course or to cower under one's cinema seat when the image of a train tugs toward the front of the screen. We do not do these things now, and we do not regard characters as sentient beings to whom one "relates" in a social sense.

There are strong grounds, however, for believing that these assumptions have no basis at all in the reality of human mental functioning. Over the past three decades, researchers in the fields of cognitive psychology, social cognition, and neuroscience have developed a strong challenge to the view of the absolute primacy of 'intellect" over "instinct" in higher-level cognitive processes, postulating an alternate view of what organizes and drives human thought. Two basic claims make up their challenge. First, human beings' conscious judgments (verbally reportable, controllable, intentional) are deeply influenced and affected by stimulus processing that takes place unconsciously (immediately, unintentionally, inaccessibly to normal awareness).[14] Second, the activity of the unconscious consists of generating *actional* impulses—impulses that effect changes in the body and have direct behavioral consequences—which precede awareness and the experience of reflection. To restate these two claims: (1) the unconscious informs and influences the conscious mind and (2) action precedes reflection. These ideas are not new. William James, followed by various cognitive theorists throughout the twentieth century, proposed the unconscious mind as the source of impulses to action, describing the role of consciousness rather as gatekeeper or sense-maker of these impulses.[15] But far clearer descriptions have emerged in recent decades of where precisely these impulses come from: from our evolved motives and preferences, from past experiences in similar situations, and from information in our social environment.[16] These things modify our conduct and cognition through contextual *priming*. In cognitive psychology, "priming" means coming into contact with stimuli, consciously or unconsciously processed, that influence behavior and choices directly. In contextual priming, the mere presence of certain stimuli in the environment automatically activates stored mental representations of those stimuli and concomitantly all of the internal information (knowledge, affect, expectations) stored in those representations that is relevant to responding back.[17]

Such priming effects depend on close, automatic connections between perception and action, and contemporary cognitive-psychological research persuasively demonstrates that physical experience is implicated in all manner of cognitive activity. Among the most important implications of these findings is that the automatic judgments we make of others have immediate behavioral correlates. When we perceive in someone a trait consistent with our knowledge of trustworthiness (such as a calming smile), for instance, the perception will "activate" a series of sensory and physical conditions consistent with our experiences of trustworthiness, providing a conceptual template to guide our emerging assessment of the person while priming and encouraging behaviors consistent with the experience of trust.[18] Observing a demeanor consistent with trust, we might automatically incline closer to the person or find ourselves willing or wishing to confide in her or him. The trusted person's behavior toward us will also be affected in turn.[19] Put simply, our judgments have a behavioral dimension, and this behavior is social or interactional by design.

A further premise of this view of cognition is that our *embodied* experience grounds our *conceptual* knowledge. Consider how references to warmth and coldness pervade our descriptions of persons. We give and receive warm welcomes, warm up to others, feel we have been given a cold shoulder, are unnerved by "cold fish." We literally perform these metaphors in social life too, putting on the kettle and lighting candles when guests drop in. And we know instinctively what these terms indicate: a cold person is self-centered, competitive, and untrustworthy, while a warm one is prosocial, generous, and trusting. Warmth and affection are synonymous. Social psychologists have been able to demonstrate the importance of "temperature"-type assessments for some time, both in forming impressions and as a main dimension of identified "out-groups" around the world. Solomon Asch's classic work on first impressions and later research on stereotyping processes both conclude that forming impressions of new acquaintances begins with an assessment of interpersonal warmth or coldness.[20] Why, though, do we use the terms *warm* and *cold* to refer to these two basic categories of persons? Asch's explanation for the power of the warm-cold dimension in interpersonal perception was that abstract psychological concepts, such as interpersonal warmth, are metaphorically based on concrete physical experiences.[21] Asch anticipated modern cognitive-psychological researchers, such as George Lakoff and Lawrence Barsalou, in "grounding" abstract concepts in perceptual and physical experience. Concepts concerning the physical world (temperature, distance, size, and other physical phenomena) form early in childhood, as they are based on direct physical experience and do not depend on higher-level language- or memory-retrieval skills.[22] According to one approach, the "conceptual scaffolding" model, abstract concepts then develop based on these physical concepts, becoming so strongly associated with them that they come to share key features.[23] Barsalou's "perceptual symbol systems" theory similarly holds that cognition is based on modality-specific, sensory-motor symbols for perception and action, and thus that mental activity is based inextricably on physical experience.[24] These theories share significant features with Lakoff and Mark Johnson's classic work on cognitive metaphor, which argues that abstract concepts "emerge" from the experience of traveling through space.[25] The close association between mental and physical functioning helps to explain the fact that humans readily and fluently use physical terms in connection with more abstract phenomena, especially social and psychological phenomena: a "close" relationship, "warm" smile, "higher calling," and so on.

It makes intuitive sense that early childhood experiences with caretakers might engender a strong associative connection between the concepts of physical and psychological warmth.[26] But much evidence now exists to support this intuition, corroborating the claim that the physical experience of warmth or coldness produces the same subjective feeling associated with these states—and vice versa. The psychologist Harry Harlow staged a famous experiment in 1958, wherein he raised infant monkeys with a food-dispensing "mother" doll made either of wire or of cloth backlit by a 100-watt bulb. These experiments were the first to demonstrate that physical warmth could effectively (in monkeys) substitute for the absence of maternal warmth, leading to significantly greater social warmth capacities for the monkeys later in adulthood.[27] More recently, Lawrence Williams and John Bargh showed that incidental experiences of warmth produce feelings of trust and behavioral effects on generosity without the person's awareness. In one of their experiments, test participants who briefly handled a cup of hot coffee and then were given a questionnaire asking them to evaluate an imaginary person regarded the person as "warmer" than did the participants who handled a cup of cold coffee in the same scenario. This neatly replicates the effects of the words *warm* and *cold* in Asch's original impression-formation study. In another, subjects touched either a hot or a cold therapeutic pad, gave the pad a product rating, and then were presented with a study participation reward that they could either keep or give to a friend. They were far more likely to choose the selfish option (keeping the reward) if they had touched the cold pad and the selfless option if they had been exposed to heat.[28] In both cases, physical warmth or coldness led to judgments and behavior that were socially warm or cold. Zhong and Leonardelli's 2008 work reversed the causal direction, showing that after an actual or remembered experience of social rejection, test participants reported the temperature of the testing room as being colder than those who had experience or recalled an experience of social warmth.[29] IJzerman and Semin further found that social proximity manipulations also produced changes in perceived room temperature.[30] Utilizing the well-known psychological phenomenon wherein similarity (even conceived in terms of physical proximity) is associated with attraction,[31] these studies show that experiences of social warmth produce concomitant feelings of physical warmth.

To experience and understand "warmth," all this research finally suggests, involves instantaneous interactions of the body, brain, and environment. Representations of the concept, which we are accustomed to locating in one domain—physical or psychological—in fact activate simultaneously across both. Crucially, at no point in any of these studies did participants show explicit awareness of the effect of physical temperature on their feelings of generosity or social connectedness or vice versa. Research has been done on humans' capacity for explicit awareness of the physical-social warmth connection and other mental processes wherein test participants who engaged explicitly with variables pertaining to temperature and social inclusion were invited to guess the experiment's hypothesis. These test results and others like them strongly support the hypothesis that people lack explicit awareness of the effects described here.[32] The mechanisms that generate our understanding of temperature are implicit, automatic—and inevitable. It is worth reiterating that in neither the research described nor the present argument is the term *automatic* employed as if it meant *determining* or *ultimate*. To insist that our engagement with and understanding of temperature is *limited* to the immediate, unintentional, or automatically supplied association of warmth with affection and trust would plainly be wrong. Were this the case, all cold experiences would be unpleasant (anybody who harbors warm feelings for ice cream or ice-cold beer would beg to differ), and there would be far more uniformity of social opinion within groups of acquaintances. There would, further, be no possibility of historically or culturally constructed perceptions of temperature and its implications, though this is not the case: Mary Floyd-Wilson and others (for instance) have investigated "climate theory"—theories of the influence of temperature on human temperament—as a distinct branch of early modern natural philosophy.[33] Barsalou and other cognitivists describe cognition as situated in particular environments and acutely context-sensitive, regarding these as core features of cognitive processing. Indeed, their theories proceed from the assumption that the varieties of human experience, thence of varied and idiosyncratic conceptualizations of any given stimulus, are potentially infinite.[34] Experience, expertise, and idiosyncrasy are not diminished by the fact of automaticity. Nor, though, can these phenomena render the influence of automatic processes obsolete.

That is, it would just as plainly be wrong, considering the evidence discussed, to deny that the concepts of warmth and affection are ineluctably connected, that humans *do* utilize heat representations in understanding other humans. Thus, when a person anticipated to be "warm" projects attributes conceptually linked to "coldness," the perceiver must exert unusual cognitive effort to comprehend this person. To sustain the preconception of him or her as "warm," the perceiver must make the sustained mental effort to counter the suggestions of context and intellect. This effort may not completely succeed; perhaps the perceiver will come to associate the object of perception, to some degree, with a sense of strain, difficulty, or disappointment. This is the typical viewer experience of Prince Hal, I contend, and this is the means by which the role consistently foils the efforts of actors to make him into the character he is presumed to be.

LUKEWARM RECEPTIONS

Ella's opinion of Hal in fact has plenty of precedent in the mainstream critical tradition. Indeed, archived reactions to productions of *1 Henry IV* attest to a long history of critical disappointment in the prince, the only real difference between Ella's and the professional reviewers' opinions being the full conviction with which the latter visit the flaws of the character upon the actors who portray him. Although only a sampling of critical excerpts appears here, they reflect a clear pattern of critical reception—namely, though there are some exceptional cases (e.g. in the rare case of an unfunny Falstaff, such as Orson Welles's opposite Keith Baxter's Hal in *Chimes at Midnight*; or of a seriously miscast Hotspur[35]), as a rule, nobody likes Prince Hal. The following excerpts come from reviews of *1 Henry IV* productions staged by the Royal Shakespeare Company since 1931:

> Eric Maxon fell a little short as Henry Prince of Wales in the metamorphosis from the pleasure seeker to the man of war. He was hardly full-blooded enough to suit Shakespeare in heroic mood. (*Birmingham Evening Dispatch*, 4.6.1931)

> Richard Burton's Prince Hal grew well enough from the prodigal youth to Hotspur's vanquisher, but he was too melancholy in his early pleasures . . . and while he found the passion and the poetry of the future king, he was lacking in regal gait and movement. (*Western Daily Press*, 5.4.1951)

> Terence Longdon has moments when he is rather less "at home" in this world of fiery spirits, gallant hearts and meddlesome women—though not because he lacks the necessary suggestion of valour and courage. He is not yet fully acquainted, one feels, with the Shakespearean theatre. (*Birmingham Weekly Post*, 12.12.1952)

> Gerard Murphy's Hal seemed to be stuck on one note of bouncy irresponsibility . . . his philandering with the lower classes never reveals the instincts of majesty that lie beneath the surface. (*Evening Standard*, 10.6.1982)

> Hal's transition from reckless youth to high-minded, duty-bound sovereign is strangely muted by William Houston's somber performance. Flinty-faced, with a monkish pudding-bowl hair-cut, Houston looks about as comfortable as a vicar at a rave when he enters Hal's old haunt. (*The Independent*, 21.2.2001)

How can it be that performances of a character with such evidently positive attributes, with folk history squarely on his side, draw consistently tepid receptions? By this point the reader may anticipate the conclusion toward which this chapter tends: that Hal is typically received coldly by critics because Hal, in the terms hitherto explored, is cold. When a critic castigates an actor for failing to deliver the Hal he imagines, the real offense he claims is the actor's failing to make Hal more deserving of the love the critic cannot feel for him. The means by which Shakespeare's text makes Hal essentially unlikeable might be summarized as follows: *1 Henry IV*, a play of diverse, divided affections and loyalties, has a relatively small cast full of conspicuously "warm" and "cold" characters. Among these characters, Hal is "cold" without question, deflecting rather than inviting the affections of audiences.

The remaining paragraphs employ the cognitive material hitherto discussed to explore *where* and *how* the play-text of *1 Henry IV* constructs these definitively social personal attributes for its characters. While no scholarly work to my knowledge has yet paid detailed attention to the linguistic and conceptual relevance of temperature to *1 Henry IV*, references and sensitivity to temperature are not absent from the scholarly tradition surrounding this play. They are, in fact, abundant, to the extent that they suggest themselves to scholars' unconscious minds by virtue of their incidental density. Two brief but representative quotes follow; the repetitious discourse has not to my knowledge been acknowledged in criticism elsewhere:

> As a public justification, this [5.4.50–56] is well enough; but as the attitude of a son to his father, who has just expressed his recovered faith in him, it is—to say the least—somewhat frigid.[36]

> [The actors who play Hotspur] make us aware of his warmth, exuberance, and poetry, qualities somewhat lacking in Hal; and they almost convince us that the rebel is more attractive than the prince. For if Hal's poise accentuates Hotspur's recklessness, Hotspur's frankness calls attention to what is calculating and disingenuous in Hal's relationships with others. . . . [Hal is] a character whose personal defects are part of his princely perfection, whose coldness and calculation are seen by Shakespeare as necessary virtues in a leader.[37]

Note too how frequently such language carries over into theatre reviews. The quotes from the RSC reviews transcribed above tell of Hals who would certainly rate low on the scale of social warmth, but critics are frequently more explicit. For instance, the *Oxford Companion to Shakespeare* entry on *1 Henry IV* mentions Michael Pennington's "cold Harry" in the English Shakespeare Company's inaugural 1986 production—the same production, somewhat curiously, that directly acknowledged Hal's folk-hero status, opening the play with a troupe of actors tramping onstage to sing a galvanizing fictional ballad of the future king.[38] Furthermore, any of Hal's intimations toward rejecting Falstaff, especially "I do, I will" (2.4.481), are described almost out of hand as "chilling."

What could account for this consistency of response, and the consistency of the terms in which these responses are conceptualized? One way to think about theatrical spectatorship is to envision spectators as rather like cognitive-psychology testing subjects, in the sense that their judgment and behavior is susceptible to unintentional influence in the form of contextual priming. To pursue the analogy, plays and psychology experiments are different from life as lived, if not in the cognitive and perceptual mechanisms required to comprehend them (for the aim and perceived value of such experiments is precisely to replicate "normal" cognitive conditions), then certainly in their degree of contrivance. Plays are constructs that produce results: shared awareness, shared emotions, shared convictions. Shakespeare need not have intended or anticipated the precise shared experiences his plays engender, but their canonicity and adaptability across historical and cultural settings surely attests to the efficacy with which they *do* generate such experiences. It is on this matter of efficacy that the comparison of plays to controlled experiments might be most productive. Cognitive studies give researchers exceptional control over the environmental factors affecting their subjects: the level of trust Sarah feels in Joe is hypothesized to be a function of her contact with heat, and researchers design tests to reveal whether the heat variable is alone accountable for her change of feeling.

Shakespeare's plays make themselves understood to audiences by comparable means. Play-texts are made of language, and this language includes words and ideas the meanings of which resonate in both physical and psychological-conceptual dimensions. Our apprehension of these words occasionally takes place in full consciousness: we are unlikely (and not expected) to "miss" the fact that Hotspur happens to be a hothead, for instance. But the influence of such words goes much farther than generally acknowledged, for they are also processed automatically, engaging the sensory regions of the brain to formulate impressions in the form of instinctive preferences and attitudes, which sustain themselves across the character's subsequent appearances. Acknowledging all this, we are well prepared to observe (and to grasp some of the cognitive and interpretive consequences) that representations of Prince Hal, and indeed of all of *1 Henry IV*'s principal male roles, are deeply informed by the physical and psychological implications of temperature. A conspicuous incidence of temperature-language spreads across *1 Henry IV* like a satellite weather system, concentrating "warmth" around some characters—specifically Hotspur and Falstaff—and "cold' around others—specifically, Hal and his father. These concepts, in turn, exert a profound influence on the level or limits of affection these characters might hope to earn from audiences. In other words, language pertaining to "warmth" functions as a linguistic behavioral prime. It therefore both reflects the affection these characters reliably earn from audiences and bears real responsibility for engendering it.

Where can we find this language? Chronology is important in assessing the efficacy of primes (since earlier impressions of characters condition subsequent ones), so it makes good sense to start with the play's very first lines: "So shaken as we are, so wan with care," spoken by King Henry. The king's words are so often understood to refer to his personal health that he is "conventionally played as a cold thin-lipped figure half way to the mortuary at curtain rise"[39]—despite clearly referring in context to the recent chaos in the kingdom, and despite there being no further reports in this play of Henry's poor health—in fact, it is hot-blooded Hotspur's father Northumberland who is later waylaid by illness. The king's opening sentiments further anticipate his troubled admission of his unnatural fondness for Northumberland's Harry over his own: "Yea, there thou make'st me sad, and make'st me sin/ In envy" (1.1.77–78). The echo effects an elision between an audience's sense of the frailty of Henry's realm and the frailty of his affections for Hal, both structured via a metaphor of bodily illness. The familiar view that *1 Henry IV* is about father-son relationships[40] rarely acknowledges how deeply and consistently the play-text engages the language and concept of affection, freely and unconditionally given or coldly, shrewdly pursued.

Outside Henry's court, one finds many depictions of contrastingly warm and mutual fellow feeling. The Rebels' quasi-familial arguments in 1.3 and 3.1 are, in this respect, conceptually of a piece with the domestic scenes featuring their wives. Evidence from the performance record indicates that these scenes contribute significantly to Hotspur's perceived warmth relative to Hal's—or at any rate, the way in which *1 Henry IV* naturally allocates warmth among its characters can easily come unbalanced by Hotspurs who neglect their marriages. The most consistently praised Prince Hal of recent years, Jamie Parker's for the 2011 season at Shakespeare's Globe, siphoned up the large helping of sympathy that the production's Hotspur, played by Sam Crane, was unable to claim as his own.[41] The reviewer for *Shakespeare Bulletin* wrote:

> Never allowing himself to indulge in affection, Hotspur was quite willing to hurt his wife privately and publicly: shunning her at their estate in England, his coldness towards his partner survived the trip to Wales and Kate was left visibly discomforted by the amorous Mortimers and tearfully regretting her refusal to sing at Hotspur's request. Any sympathy that might have attached to Hotspur as a result of a caring domestic partnership was thwarted.[42]

The lesson here is that a Hotspur without charm and warmth presents Hal with chances that the play ordinarily denies him. (Most of the time, though, Hotspur has charm.) The affection-warmth conceptual network concentrates itself around the Eastcheap contingent as well. One instance is Falstaff's comparison of Bardolph's nose to a torch-lamp, in "O, thou art a perpetual triumph, an everlasting bonfire-light!" (3.3.42–50). Indeed, an important feature of Falstaff's characterization is the depth of affection and trust he places in his companions. His loyalty to and trust in Hal—even as shown in incidental matters, like his request for help at Gad's Hill, "I prithee, good Prince Hal, help me to my horse, good king's son" (2.2.40–41)—can be heartbreaking. But note too, referring to Poins: "I have forsworn his company any time this two and twenty years, and yet I am bewitched with the rogue's company. . . . If the rascal have not given me medicines to make me love him, I'll be hanged" (2.2.15–19).

Consider, too, Hal's notorious speech in Scene 1.2, beginning "I know you all":

> Yet herein will I imitate the sun
> Who doth permit the base contagious clouds
> To smother up his beauty from the world,
> That, when he please again to be himself,
> Being wanted, he may be more wond'red at,
> By breaking through the foul and ugly mists
> Of vapors that did seem to strangle him. (1.2.201–207)

The perspectives of automatic and embodied cognition equip us handily to perceive how this speech swiftly undoes what it appears to set out to do for Hal. Hal aims here to communicate two things: that the persona he has presented thus far in the play is "not the real me"—he may have identified earlier in the scene with the "moon's men" but is actually a virtuous sort who, sunlike, "goes by Phoebus"—and that he intends to use his degraded reputation to political advantage, staging a premeditated rise into political agency as if by miraculous redemption. However, Hal's chosen central metaphor of the sun emerging from clouds undermines the credibility of both these claims, for he manages to associate himself more strongly with the concepts of smothering, clouds, and mist than with warming, positive sunshine. These operations are prepared for by the preceding dialogue, as Hal's and Falstaff's talk of the moon "under whose countenance we steal," and Poins's provision of masks ("our vizzards we will change") and coarse cloth to "immask our noted outward garments" (1.2.180–184) set darkness and concealment into conceptual play, ready to cluster around the prince. The perceptual processing mechanisms discussed above entail that an audience's understanding of Hal's words involves automatic, instantaneous bodily enactment—meaning that, as Hal's sun is smothered in cloud, his words literally have cooling effect.

Furthermore, Hal's "I know you all" is most obviously understood as a verbal gesture of familiarity, after which audiences reasonably expect disclosure, insights into Hal's hitherto ambivalent identity. The message we actually receive, though, is a variation on Iago's "I am not what I am" (*Othello* 1.1.65). Hal describes his present condition and alludes to future developments but denies his hearers insight into the subject who stands and speaks a soliloquy (his only one in the play). The speech may point to Hal's secret intention to reveal himself soon as the ruler of legend, but it never makes his appealing qualities available, through behavior or language, to sensory or psychosocial perception. From an embodied-cognitive perspective, it never makes them "real." Soliloquys are fundamentally generous modes of address, means of generating warmth and rapport between character and audience; but this soliloquy holds its subject at a disquieting remove. The audience, in other words, is distanced from Hal, and the play maintains this distance. The Gad's Hill robbery scene is a case in point. When the Prince and Poins emerge from hiding to find the travelers successfully robbed and bound by Peto, Gadshill, and Falstaff, Hal declares, "Now could thou and I rob the thieves and go merrily to London, it would be argument for a week, laughter for a month, and a good jest forever" (2.2.93–96). A few moments later, the pair duly undertake precisely what Hal said they would. Their successful operation leaves Hal in a merry mood, and he comments in hysterics:

> Got with much ease. Now merrily to horse. The thieves are all scattered, and possessed with fear so strongly that they dare not meet each other: . . . Falstaff sweats to death and lards the lean earth as he walks along. Were't not for laughing, I should pity him. (2.2.105–111)

Hal is having great fun. It is not readily apparent, however, that the scene is (or possibly could be) entertaining to anybody but himself. Fear, frantic scattering, sweating: it is not inconceivable that these could ever be sources of amusement, but the text offers little guidance on or opportunity for making them so. Directors sometimes use the sketchiness of the stage directions as grounds for contriving elaborate action sequences.[43] But such scenes, however effectively diverting, are not licensed by a script that has the thieves *all run away* more or less instantly, with Falstaff following suit *after a blow or two* (2.2.101–104); the prince summarizes the action thus: "Got with much ease."[44] Never much of a favorite with directors or audiences, the Gad's Hill robbery routinely disappoints in performance, in much the same way that Hal does: both never really deliver what they promise. This is especially conspicuous in light of the tavern scene that follows, where Falstaff provides a revisionist history of the robbery and is caught in the lie. This scene never misses: indeed, it affords the actor playing Falstaff as sure a triumph in the audience's eyes as the *Pyramus and Thisbe* performance in *A Midsummer Night's Dream* affords the lucky actor playing Flute the Tinker. Note too how Poins's proposal of the robbery to Hal looks forward to the latter scene (2.4), not the robbery itself, for its promising comic payoff: "The virtue of this jest will be the incomprehensible lies that this same fat rogue will tell us when we meet at supper: how thirty, at least, he fought with; what wards, what blows, what extremities he endured; and in the reproof of this lies the jest" (1.2.189–194). The Gad's Hill robbery is a private, quite unfathomable victory for Hal; its retelling, a public and broadly inclusive victory for Falstaff.

Where Hal's purported radiance is frustratingly obscured, the character of Hotspur does exactly what it says on the tin. The aptness of Henry Percy's historical nickname in light of his character is remarked on in the play (5.2.16–19) and presumably was never far from audiences' awareness. It is a fine detail that Percy himself senses the potential of metaphors to sway perceptions: when Vernon describes Hal and his allies armed for Shrewsbury as "Glittering in golden coats like images;/ As full of spirit as the month of May/ And gorgeous as the sun at midsummer" (4.1.99–101), Hotspur cuts him short with "No more, no more! Worse than the sun in March,/ This praise doth nourish agues" (4.1.110–111). Curtailing the power of Hal's reported sunlike gorgeousness with his own image of steely, chills-inducing spring sunlight, Hotspur proceeds to elaborate an impressive opposing form of metaphorical heat: "to the fire-eyed maid of smoky war/ All hot and bleeding will we offer them!. . . I am on fire/ To hear this rich reprisal is so nigh,/ And yet not ours" (4.1.113–18). This outburst challenges Vernon's praise in a contest of "heat" metaphors, for which the prize is the audience's confidence in one or the other Harry's triumph in battle.

And when Hal actually *does* triumph at Shrewsbury? Shakespeare neither denies nor begrudges Hal the attainment of his destiny. Nevertheless it is worth pointing out that the glorious climax wherein *the Prince killeth Percy* provides the actors of both Hotspur and Falstaff with some of their finest opportunities to charm and delight. Hotspur's death speech—"Oh Harry, thou has robbed me of my youth!" (5.4.76–85)—is a perfect vehicle for actors to draw upon and concentrate all their charisma and best inventions for the role. This is perhaps why reviewers of *1 Henry IV* productions so often mention the actor's handling of this speech. And, while there is no historical survey of Shakespeare's best-loved theatrical moments, it surely would surprise no one if—were such a list to materialize one day—the moment where *Falstaff riseth up* (5.4.109) stood near the top.

LASTING IMPRESSIONS

This chapter began with the hypothesis that audience members perceive *1 Henry IV*'s Prince Hal as "cold." An extended inquiry into body-based conceptual processing enabled us to identify the strategy of character development used to create this version of the madcap prince. Shakespeare creates a verbal matrix around Hal, consisting of concepts (here, warmth/coldness) that promote interpersonal impressions automatically via the mechanisms discussed. Simultaneously, he sets these impressions *against* the understanding of the character that "higher" cognitive analysis—historical or literary precedent, genre convention, and so on—makes available. He thus continuously forces a conflict at the level of reception between what his audiences know they should feel about the personality presented to them and what they actually do feel. The quality of interest Shakespeare's audiences take in Hal is made complex and exciting by the conflicts the play creates between reflection and action, conscious appraisal and automatic response. The phenomenon of Hal's coldness ultimately supports the hypothesis of the embodiment of human mental and social life. But the question of *why* persists. Why did Shakespeare craft a representation of Hal that slyly defies the guidance of genre, advising of popular political theory, and template of history and legend? Why pique audiences' intuitions that the future Henry V is really a cold fish? One might consider that Hal's characterization reconciles well with a view of *1 Henry IV* as a play focused on the topics of personal and political allegiance, and with the qualities that conspire to generate them—qualities both logical and intangible, subject to analysis and accessible only to instinct. Hal's peculiar detachment, which Derek Traversi described long ago as his great political virtue,[45] proceeds from the same source that distances him from the affections of a public all but conscripted to love him. Without offering political commentary, then, Shakespeare provides a drama about political opinion—what stuff 'tis made of, whereof it is born. Perhaps he recognized that in political affairs, spectators' impressions of the performers' impalpable social qualities—however engendered, however justified—are of incalculable consequence; and he may have recognized the theatre, so similar in its mechanisms of inspiring and holding assent, as the right medium to treat this complex theme.

NOTES

1. All Shakespeare references are to G. Blakemore Evans and J.J.M. Tobin, eds., *The Riverside Shakespeare*, 2nd ed. (New York: Houghton Mifflin, 1997). All line numbers refer to *The First Part of Henry IV* unless otherwise stated. The completion of this chapter was in part made possible through the generosity of The Shakespeare Birthplace Trust, who funded my attendance at the World Shakespeare Congress in Prague, July 2011.
2. Henry V's heroic nature was popularized in ballads such as the "Agincourt Carol" (Trinity Carol role) and "King Henry Fifth's Conquest of France"; see John Jacob Niles, *The Ballad Book of John Jacob Niles* (Lexington: University Press of Kentucky, 2000), 264–269.
3. Samuel Johnson, "The Life of Dryden," in *Lives of the English Poets*, ed. G.B. Hill, 3 vols. (Oxford: Clarendon Press, 1905), vol. 1, 403.
4. William Archer, "The Ethics of Theatre Criticism," in *About the Theatre: Essays and Studies* (London: T. Fisher Unwin, 1886). Accessed January 4, 2013. http://www.theatredatabase.com/20th_century/ethics_dramatic_criticism.html.
5. Gerald Bordman and Thomas S. Hischak, "Drama Criticism in America," in *The Oxford Companion to American Theatre* (Oxford: Oxford University Press, 1984), 186.
6. Samuel C. Carpenter, *The Mirror of Taste*, and http://archive.org/details/themirroroftaste22488gut. Italics are mine.

7. Harold Clurman, *The Collected Works of Harold Clurman*, ed. Marjorie Loggia and Glenn Young (New York: Applause, 1994), vi.
8. Dwight Garner, *The New York Times*, August 19, 2012.
9. Stephen Booth, "Poetic Richness: A Preliminary Audit," *Pacific Coast Philology* 19.1/2 (1984): 68–78; esp. 68. See also "The Function of Criticism at the Present Time and All Others," *Shakespeare Quarterly* 41.2 (1990): 262–268.
10. Booth, "Function," 263.
11. For the disfavor of character criticism in academic circles in the second half of the twentieth century, see Paul Yachnin and Jessica Slights, introduction to *Shakespeare and Character: Theory, History, Performance and Theatrical Persons*, ed. Yachnin and Slights (New York: Palgrave MacMillan, 2009): 1–19.
12. William Hazlitt, *Characters of Shakespeare's Plays* (Oxford: Oxford World Classics, 1917), 104.
13. See Edwin A. Locke and Gary P. Latham, "Building a Practically Useful Theory of Goal Setting and Task Performance: A 35 Year Odyssey," *American Psychologist* 57 (2002): 705–717, a review of 35 years of research of empirical research on goal-setting theory, for evidence of the persistence of these assumptions in mainstream psychology.
14. For the past quarter-century "unconscious" phenomena have largely been conceptualized as "unintentional." See Richard E. Nisbett and Timothy DeCamp Wilson, "Telling More Than We Can Know: Verbal Reports on Mental Processes," *Psychological Review* 84 (1977): 231–259, which posed the question "To what extent are people aware of and able to report on the true causes of their behavior?" and answered, controversially at the time, "not very well."
15. William James, *Principles of Psychology*, vol. 2 (New York: Holt, 1890); Michael Gazzaniga, *The Social Brain* (New York: Basic Books, 1985); Benjamin Libet, "Unconscious Cerebral Initiative and the Role of Conscious Will in Voluntary Action," *Behavioral and Brain Sciences* 8 (1986): 529–556; Daniel Wegner, *The Illusion of Conscious Will* (Cambridge, Massachusetts: MIT Press, 2000).
16. See the work of Antonio Damasio, beginning with "The Somatic Marker Hypothesis and the Possible Functions of the Prefrontal Cortex," *Philosophical Transactions of the Royal Society of London* Series B: Biological Sciences, 351 (1996): 1413–1420; and Norbert Schwarz and Gerald L. Clore, "Feelings and Phenomenal Experiences," in *Social Psychology*, ed. Edward T. Higgins and Arie W. Kruglanski (New York: Guilford Press, 1996): 433–465.
17. Comprehensively described first in Edward T. Higgins and John A. Bargh, "Social Perception and Social Cognition," *Annual Review of Psychology* 38 (1987): 369–425.
18. For a straightforward review of how such conceptual processing occurs, see Lawrence W. Barsalou, W. Kyle Simmons, Aron K. Barbey, and Christine D. Wilson, "Grounding Conceptual Knowledge In Modality-Specific Systems," *Trends in Cognitive Science* 7.2 (2003): 84–91.
19. See Ap Dijksterhuis and John A. Bargh, "The Perception-Behavior Expressway: Automatic Effects of Social Perception on Social Behavior," in *Advances in Experimental Social Psychology 33*, ed. Mark P. Zanna (San Diego, California: Academic Press, 2001): 1–40.
20. Solomon Asch, "Forming Impressions of Personality," *Journal of Abnormal and Social Psychology* 41 (1946): 258–290; Susan T. Fiske, Amy J. C. Cuddy, Peter Glick, and Jun Xu, "A Model of (Often Mixed) Stereotype Content: Competence and Warmth Respectively Follow from Perceived Social

Status and Competition," *Journal of Personality and Social Psychology* 82 (2002): 878–902. Another relevant classic in social psychology is Harold H. Kelley, "The Warm-Cold Variable in First Impressions of Persons," *Journal of Abnormal and Social Psychology* 46 (1951): 457–469.

21. Solomon Asch, "The Metaphor: A Psychological Inquiry," in *Person Perception and Interpersonal Behavior*, ed. Renato Tagiuri and Luigi Petrullo (Stanford, California: Stanford University Press, 1958): 86–94.
22. See Jean M. Mandler, 'How to Build a Baby II: Conceptual Primitives,' *Psychological Review* 99 (1992): 587–604.
23. Lawrence E. Williams, Julie Y. Huang, and John A. Bargh, "The Scaffolded Mind: Higher Mental Processes Are Grounded in Early Experience of the Physical World," *European Journal of Social Psychology* 39 (2009): 1257–1267.
24. Lawrence W. Barsalou, "Perceptual Symbol Systems," *Behavioral and Brain Sciences* 22 (1999): 577–660; and Barsalou, "Grounding," 84–91.
25. George Lakoff and Mark Johnson, *Metaphors We Live By* (Chicago: University of Chicago Press, 1980).
26. Lakoff and Johnson also defend the "primary metaphor" status of "affection is warmth" by citing the coexistence of physical and psychological warmth in early life; see their *Philosophy in the Flesh: The Embodied Mind and Its Challenge to Western Thought* (New York: Harper-Collins, 1999): 45–60.
27. Harry F. Harlow, "The Nature of Love," *American Psychologist* 13 (1958): 673–685.
28. Lawrence E. Williams and John A. Bargh, "Experiencing Physical Warmth Promotes Interpersonal Warmth," *Science* 322 (2008): 606–607.
29. Chen-Bo Zhong and Geoffrey J. Leonardelli, "Cold and Lonely: Does Social Exclusion Literally Feel Cold?," *Psychological Science* 19 (2008): 838–842.
30. Hans IJzerman and Gün R. Semin, "Temperature Perceptions as a Ground for Social Proximity," *Journal of Experimental Social Psychology* 46 (2010): 867–873.
31. See Donn E. Byrne, *The Attraction Paradigm* (New York: Academic Press, 1971).
32. See Richard E. Nisbett and Nancy Bellows, "Verbal Reports about Causal Influences on Social Judgments: Private Access vs. Public Theories," *Journal of Personality and Social Psychology* 35 (1977): 613–624.
33. Mary Floyd-Wilson, "Temperature, Temperance, and Racial Difference in Ben Jonson's *The Masque of Blackness*," *English Literary Renaissance* 28.2 (1998): 183–209.
34. See Lawrence W. Barsalou, Christine D. Wilson, and Wendy Hasenkamp, "On the Vices of Nominalization and the Virtues of Contextualizing," in *The Mind in Context*, ed. Batja Mesquita (New York: Guilford Press, 2010): 334–360.
35. In the Shakespeare's Globe production, Summer 2011, Jamie Parker's Hal was widely praised, while Sam Crane's Hotspur was reedy, nervous, and difficult to hear. Carol Chillington Rutter, "Shakespeare Performances in England 2010," *Shakespeare Survey* 64 (2011): 340–377; esp. 344, memorably described this Hotspur as "his own caricature, a gesticulating human parenthesis in doublet and hose with an adenoidal voice and strange tea-kettle lisp."
36. Derek Traversi, *Shakespeare from Richard II to Henry V* (Stanford, California: Stanford University Press, 1957), 103.
37. Robert Ornstein, *A Kingdom for a Stage: The Achievement of Shakespeare's History Plays* (Cambridge, Massachusetts: Harvard University Press, 1972), 134–135.

38. Scott McMillin, *Henry IV Part I*, Shakespeare in Performance (Manchester: Manchester University Press, 1991), 106.
39. Michael Billington, "Review: *1 Henry IV*," *The Guardian*, April 22, 1975.
40. See Coppélia Kahn, *Man's Estate: Masculine Identity in Shakespeare* (Berkeley: University of California Press, 1981), 70–74.
41. See Rutter, cited previously.
42. Laura Grace Godwin, "Henry IV, Part 1 and Part 2 (Review)," *Shakespeare Bulletin* 29.1 (2011): 105.
43. As, for instance, in the 2011 staging at Shakespeare's Globe, which had Roger Allam's Falstaff flailing madly at the stage's wooden scaffold. Godwin, 105.
44. Compare the masque in *The Tempest* or *The Winter's Tale*'s dance of shepherds and shepherdesses, both of which are acknowledged as having commanded a decent interval of stage time.
45. Traversi, 103.

[. . . First Link . . .]

Subjectivity and the Mind-Body: Extending the Self on the Renaissance Stage

Garrett A. Sullivan, Jr.

Near the end of their seminal essay "The Extended Mind," Andy Clark and David J. Chalmers ask "What, finally, of the self? Does the extended mind imply an extended self?" Their tentative answer: "It seems so." The authors substantiate their conclusion by referencing Otto's now famous notebook as "a central part of his identity as a cognitive agent" and "Otto *himself* . . . as an extended system, a coupling of biological organism and external resources."[1] The provocative claim that subjectivity can be understood as "an extended system" has not been assimilated within early modern literary criticism. In what follows, I isolate two reasons why before arguing that Renaissance drama presents us with examples of extended selfhood that pose a strong challenge to traditional understandings of subjectivity.

A major implication of work on the extended mind is that neither cognition nor embodiment end at the (porous) edges of the human body. Instead, embodiment extends beyond and across bodies and environments; its contours are different from those of the body itself. And yet scholars tend to smudge the distinction between body and embodiment by mapping the latter onto the former; that is, they approach embodiment as if the body necessarily limns the terrain of its operations. The early modern mind-body demands that we think embodiment differently—that we recognize it as constituted out of somatic transactions with the world.

Like embodiment, subjectivity has traditionally been territorialized onto the individual body. More specifically, it is commonly treated as a synonym not merely for an atomized consciousness, but also for higher-order cognitive processes occurring within skin and skull. Of course, scholars of early modern literature have often stressed the social dimensions of subjectivity.[2] In doing so, however, they have seldom doubted that it exists only in the body, if not the head.[3] Put differently, while subjectivity is routinely seen as social, it is seldom understood as transactional.

In Renaissance drama, the soliloquy appears to be the natural habitat of an atomized view of subjectivity: an individual character, often alone on stage, shares his or her thought processes with the audience. One could argue that the soliloquy is not the hallmark of individuality it appears to be, informed as it is by the diegetic exchanges that precede it as well as the extradiegetic transaction with the audience that is its major precondition for being. While this argument is persuasive (and I develop a variation on it below), my point here is a simpler one: even granting the nontransactional nature of the soliloquy, it should not be taken to represent subjectivity per se. Hamlet's soliloquies do not model an early modern understanding of subjectivity as much as they represent *one* understanding among several that are available on the Renaissance stage.

A different approach to subjectivity is on offer in the relationship between Ferdinand and Bosola in Webster's *The Duchess of Malfi*.[4] Early in the play, Antonio says of Ferdinand that "He speaks with others' tongues, and hears men's suits/ With others' ears; will seem to sleep o'th'bench/ Only to entrap offenders in their answers" (1.1.173–175). Unwittingly, Antonio captures the nature of the relationship between Ferdinand and his murderous servant Bosola. Until the two fall out over the death of the Duchess, the seemingly sleeping Ferdinand remains awake and active through Bosola; even while in Rome, Ferdinand is able to take action in Malfi. Both his somatic capacities—speaking and hearing through others—and his subjectivity are distributed across bodies as well as disparate geographic spaces. In *The Duchess of Malfi*, Webster provides us with a sinister version of the subjectivity of service, by which the selfhood of the master is extended through and across the actions of his servant. Renaissance tragedy routinely stages the breakdown of this model of subjectivity, usually through the self-interested actions of ambitious servants. But this model's breakdown should not blind us to the fact that it represents a common dramatic alternative to the vision of subjectivity enshrined in the soliloquy. Importantly, the subjectivity of service is fundamentally transactional; it assumes the extended nature of the (master's) self.

Thomas Kyd's *The Spanish Tragedy* presents us with an intriguing model of extended subjectivity predicated upon affect. Over the course of his exchanges with Revenge, the ghost of Don Andrea grows increasingly impatient with his companion's apparent inactivity. His frustration comes to a head when he discovers that Revenge has fallen asleep. Revenge urges Andrea not to fret: "Though I sleep,/ Yet is my mood soliciting their souls" (3.15.19–20). Needless to say, Andrea's desire for vengeance is eventually realized, if not entirely sated; even while sleeping, Revenge has indeed solicited the souls of Hieronimo, Isabella and Bel-imperia to his cause.

How are we to understand the nature of Revenge's agency? He has shaped the central actions of the play without directly intervening in them. More importantly, he has informed the subjectivities of the avengers by "soliciting their souls" to his mood. This is particularly so in the case of Hieronimo, who spends much of the play meditating upon the relationship of justice to revenge and, of course, upon the questionable rightness of his own claims to vengeance. What we learn from Revenge's exchange with Andrea is that Hieronimo's deliberations have been gradually conditioned by the "mood" of Revenge, the affect of anger.[5] These deliberations, as well as Hieronimo's subjectivity more generally, are represented primarily via soliloquy. And yet, through its susceptibility to Revenge's influence, the monadic model of subjectivity enshrined in the soliloquy proves to be both transactional and affectively charged, as Kyd foregrounds the centrality of emotion to cognitive processes. Moreover, affect in this play is contagious, passing from Revenge to the "souls" of many of the play's central characters. Of course, Revenge is an allegorical figure, which might lead us to discount Kyd's approach to affect and subjectivity as unrealistic. However, his approach emerges out of early modern natural philosophy, which emphasizes the shaping power of the passions, their dangerously contagious nature, and their ability to "both envelop and re- or de-constitute the human subject."[6] In other words, Revenge functions as the passions do, informing and inflecting the subjectivities of those upon whom they operate.

Kyd and Webster present us with powerful examples of the extended self that both emerge out of the early modern mind-body and challenge our preconceptions about period conceptions of subjectivity. For these writers, subjectivity is not an exclusively monadic operation. Moreover, it accommodates more than familiar acts of higher-order cognition, such as the tragic hero's musings upon mortality. Indeed, what Kyd and Webster demonstrate is that subjectivity on the Renaissance stage is a remarkably varied and variegated affair, accommodating a variety of cognitive and somatic processes—even sleep!—and extending across bodies and spaces in ways that critics have only begun to account for.

[...]

NOTES

1. This essay, first written in 1995, is reprinted in Andy Clark, *Supersizing the Mind: Embodiment, Action, and Cognitive Extension* (Oxford and New York: Oxford University Press, 2011): 220–332; esp. 232.
2. See, for example, Katharine Eisaman Maus, *Inwardness and Theater in the English Renaissance* (Chicago: University of Chicago Press, 1995).
3. A recent exception is Gail Kern Paster, "Thinking with Skulls in Holbein, *Hamlet*, Vesalius, and Fuller," *The Shakespearean International Yearbook* 11 (2011): 41–60; see also Mary Floyd-Wilson's contribution to this volume, which attests to the possibility of a diabolic influence on cognition. In such an instance, subjectivity has a cosmological dimension.
4. This paragraph and the next develop in a new direction an argument first made in Garrett A. Sullivan, Jr., *Memory and Forgetting in English Renaissance Drama: Shakespeare, Marlowe, Webster* (Cambridge: Cambridge

University Press, 2005). Quotations from *The Duchess of Malfi* and *The Spanish Tragedy* are drawn from *English Renaissance Drama: A Norton Anthology*, ed. David Bevington, Lars Engle, Katharine Eisaman Maus, and Eric Rasmussen (New York: W. W. Norton, 2002).

5. While the desire for vengeance is not identical to anger, the two are tightly connected in classical and early modern thought; see, for example, Gordon Braden, *Renaissance Tragedy and the Senecan Tradition: Anger's Privilege* (New Haven, Connecticut: Yale University Press, 1985).
6. Garrett A. Sullivan, Jr., *Sleep, Romance, and Human Embodiment: Vitality from Spenser to Milton* (Cambridge: Cambridge University Press, 2012), 47. See also Gail Kern Paster, Katherine Rowe, and Mary Floyd-Wilson, ed., *Reading the Early Modern Passions: Essays in the Cultural History of Emotion* (Philadelphia: University of Pennsylvania Press, 2004); and John Sutton, "Controlling the Passions: Passion, Memory, and the Moral Physiology of Self in Seventeenth-Century Neurophilosophy," in *The Soft Underbelly of Reason: The Passions in the Seventeenth Century*, ed. Stephen Gaukroger (London and New York: Routledge, 1998): 115–146.

4 "Some Fury Pricks Me On"

Satanic Thinking in Thomas Heywood's *A Woman Killed with Kindness*

Mary Floyd-Wilson

One of the puzzles of Thomas Heywood's domestic tragedy *A Woman Killed with Kindness* is why Anne Frankford, the ostensibly content and virtuous wife, commits adultery. She seems, as Paula McQuade has noted, to lack a psychological motive for her behavior. Indeed, at the moment that she assents to the affair, she exhibits confusion but shows no signs of sexual desire or willfulness. Provocatively McQuade argues that Anne demonstrates how an early modern wife's submissive role in marriage incapacitates her moral conscience. If her husband is absent, Anne's deference to patriarchal rule leaves her unable to make moral choices for herself.[1] But to an early modern audience, the most common and pervasive plague to any person's moral capacities would not be traditional gender roles but the infectious and insidious influence of the Devil. In keeping with reformist understandings of the Devil's agency, Heywood indicates that the cognitive processes of both Anne and her seducer, Wendoll, are affected by a literal (though not necessarily physical) diabolical presence. In our effort to understand the early modern experience of thinking with an embodied and extended pre-Cartesian mind, we need to account for the period's most pressing question about cognition: can you call your thoughts your own, or were they planted by the Devil?[2]

Scholarly accounts of supernatural and preternatural forces on the early modern stage tend to adhere to the secularization thesis. The familiar trajectory charts a development of dramatic character "from the personified abstractions" of medieval vices and devils to the "represented human beings inspired by secular observation."[3] Moving too quickly from the residual notion of the Catholic incarnate Devil with whom people knowingly made pacts to a post-Enlightenment notion of human evil within a person has left us with little understanding of how the Devil operated in England during the late sixteenth and early seventeenth centuries.[4] Many scholarly treatments of demonology transmute into discussions of disbelief or emergent skepticism, an approach epitomized by Stephen Greenblatt's influential account of the period controversies over possession and exorcism in "Shakespeare and the Exorcists."[5] Analyzing the relation between *King Lear* and Samuel Harsnett's *Declaration of Egregious Popish Imposture*,

Greenblatt argues that Shakespeare's play and the theater demystify the supernatural. By exposing an array of counterfeit possessions and exorcisms, Harsnett's pamphlet "kills the credibility of the supernatural."[6] For Greenblatt, there is a correlation between the theatricality of staged exorcisms and the emptying out of rituals enacted in the early modern theatre. But Greenblatt's modern rationalist position sets up a false binary between fraudulent assertions of demonic agency and the reality of a "disenchanted grasp of the mechanics of illusion mongering."[7] The prominent binary at work in the period had nothing to do with disenchantment: instead, the presumptuous claim of exorcists (both Catholic and Puritan) that they could cast out devils was pitted against the singularity of God's power to combat the subtle and often invisible ways the Devil acted on men. In other words, exorcists were fraudulent because they claimed abilities attributed only to God. Early modern scholars and theologians may have debated the scope of the Devil's power, but no one doubted that the Devil could affect the behavior, emotions, and actions of men and women.[8] While possession by the Devil (fraudulent or not) produced spectacularly theatrical results, much of the Protestant demonological discourse expressed greater fears about the imperceptible, subtle, and secret effects that demons wreaked.

Many of the influential historians of witchcraft invoke the secularization thesis in discussing possession.[9] Michael MacDonald, for example, in his analysis of the physician Edward Jorden's interpretation of the Mary Glover case, identifies Jorden's treatise *A briefe discourse of a disease called the suffocation of the mother* (London, 1603) as a significant text in the skeptical camp.[10] Mary Glover was the central victim in a politically controversial case of alleged bewitchment by a neighbor woman named Elizabeth Jackson. Glover's dramatic fits prompted a remarkable debate over whether she suffered from a natural disease or demonic possession. Jorden, who had represented the Church's official position against the Puritans in the controversy, contended in his treatise that many of the symptoms that looked like demonic possession were, in actuality, the effects of a natural disease called "suffocation of the mother," an affliction of the womb that later became more popularly known as hysteria.[11] There is little question that Jorden intends to downplay both witchcraft and demons as a potential cause of Glover's sufferings. However, to identify Jorden as a skeptic not only minimizes his concession that witchcraft can cause illnesses but also, more importantly, his acknowledgment that the Devil can cause natural diseases. Jorden notes that the Devil can function as "an externall cause, [and] may inflict a disease by stirring up or kindling the humors of our bodies, and then depart without supplying continuall supernaturall power unto it," thus producing a natural ailment.[12] Jorden's contemporaries regularly questioned whether "witches" had the capacity to curse or charm others, and they often doubted cases of actual demonic possession. But the Protestant Devil remained a constant threat to people's spiritual and physical well-being. In fact, it could be argued that that the secularization thesis has shrouded the ways in which the seventeenth-century Devil had a greater and more insidious reach than his demonic predecessors.

As Stuart Clark has demonstrated, demonology was central to early modern understandings of science, and the most advanced medical knowledge was no exception.[13] Even if an early modern physician disputed the possibility of demonic possession, he did not deny the Devil's capacity to disturb a person's body or mind. As the French physician Andre Valladier contends, the Devil had full power over all the spirits and humors of the body and could inflict any medical disorder on a person. The medical writer Jean Fernel devotes a portion of his discourse to explaining how devils enter the human body and generate illnesses. Henry Holland maintains that the Devil knows how to apply poisons that have hidden, unknown, and strange operations in the bodies of men. Similarly, Daniel Sennert concludes that since the Devil understands the sympathies and antipathies in nature, he can cause sickness by secret poisoning.[14]

While early modern medical discourse establishes that the Devil can produce both common and uncommon physical ailments, reformist theologians also emphasized that the Devil could affect a person's body, spirit, and mind in hidden ways much less spectacular than possession. According to demonologists, there were three basic levels of demonic affliction: possession, obsession, and temptation.[15] Possession, as Jorden observes, was rare and involved not only the Devil's usurpation of the victim's body and mind but also extravagant symptoms, including contortions, spasms, bizarre speech, and the regurgitation of strange objects, such as pins, nails, feathers, stones, and pieces of flesh.[16] Moreover, possession was a communal event that drew an audience and commentators.[17] In contrast, cases of obsession were more subtle, producing fewer manifest signs. As John Cotta explains:

> I call them obsessed, in whose bodies outwardly appearing no extraordinarie signes or tokens of the Diuels corporall presidence, or residence in them (as was in the possessed manifest) yet are their mindes, vnderstanding, wils, and reason palpably obserued to bee besieged, captiued and inchanted, by an extraordinary and more then naturall, or rather an infernall inuasion of the Diuels illusions . . .[18]

But the most common and prevalent kind of encounter that people had with the Devil was temptation. Indeed, reformist writers and ministers insisted that demonic temptation was a daily occurrence. Temptation is a process, they argue, in which the Devil "conuayes into" a person's mind "either by inward suggestion or by outward object, the motion or cogitation of that sinne which he [the Devil] would have him to commit."[19] Neither an abstract nor metaphorical experience, temptation for the reformed Christian was simultaneously a physiological, cognitive, and spiritual affliction in which the Devil planted foreign images and thoughts in one's mind.

As Nathan Johnstone has demonstrated, the Protestant Devil's capacity to instill thoughts in people's minds meant that demonic influence became more subtle and insidious during the reformation. Unlike the predominantly physical and visible Catholic Devil, the Protestant Devil could usurp a person's cognitive processes in covert and invisible ways. Even as Protestant theologians proved reticent about the "physicality . . . of diabolic experience," they augmented the Devil's scope, allowing for the "widest possible transmission" of demonic effects.[20] Early modern writers treated the Devil's power much like contagion. As William Gouge explains it, "The spirituall nature of Diuels doth many waies aggrauate their terror," for he can easily "enter into" the "bodies of men and beastes" and "make them with violence rush and runne headlong hither and thither." Demons are able to "darken mens understanding, and cause much trouble and anguish in their soule and conscience."[21] According to John Downame in *The Christian Warfare*, it is a greater struggle to "wrastle" with "enemies" not of

> flesh and blood like unto our selves but with spirituall wickednesses, which are most dangerous because being spirits they can with incredible swiftnes passe from place to place which are farre distant, and therfore the more fitly take all advantages either in assaulting us at their pleasure, or withdrawing themselves when they finde resistance; being spirits they can lie secretly in ambushment, even in our bedchambers, and so surprize us when they find us most retchlesse and secure.[22]

According to Richard Gilpin, the Devil's knowledge of people's thought processes is what makes him so dangerous and effective. He has immediate access, for example, to "Objects in our Fancy, or Phantasins." He also knows how to "dart into our Minds" his "own Suggestions, and Temptations," and he understands the "secret workings of our Passions, as Love, Desire, Fear . . . because these depend upon, or are in concomitancy of the motions of the Blood and Spirits, which he can easily discern, though their motions and workings may be kept secret from the observation of all bystanders."[23] Indeed, one's own secret and unrevealed thoughts could be the very vulnerability that invites the Devil in. While incarnate Devils proved a rarer phenomenon in the seventeenth century, demonic infection was an ever-present and invasive threat to the godly and unregenerate alike.

Notably the written accounts of demonic temptation in the period indicate that it is not only a private experience of mental and emotional anguish but also a tangible encounter between the Devil and his victim. Temptation narratives describe experiencing the Devil's presence as a force felt both externally and internally. In *The young-mans second warning-peece* (1643), for example, A. W. explains that Satan pressed himself "so vehemently on my mind, that I had no power to resist him."[24] Vavasor Powell's publication *Spirituall Experiences* (1653) presents a collection of reports that relate people's internal battles with the Devil. Repeatedly the narrators assert that the Devil has put ideas or images in their heads or compelled them to act against their own wishes. One victim, called M. W., laments,

> My spirits were almost dryed up . . . [a]nd my afflictions were doubled upon me by Satan's malice who tempted me with evil thoughts. . . . When I had prayed (with a troubled spirit as well as I could) the Devill put thoughts into my minde, quite contrary to the frame of a praying spirit.[25]

It is M. W.'s specific complaint that even when he names God in his distress, the Devil "put[s] into my minde the objects of bruite creatures." A narrator named E. R. recalls telling a lie at 11 years of age, which led to years of internal struggles with the Devil, who almost persuades the speaker to commit suicide on multiple occasions.[26] In the printed account of her death, Katherin Brettergh insists that she cannot pray because Satan continually interrupts her.[27] Tormented speakers frequently express fear that the Devil will induce them to do harm to friends and family or fetch them away to hell, but the most common anxiety is that the Devil has already contaminated their innermost thoughts.[28]

Since the Devil had the power to "enter directly into the consciousness and introduce thoughts that were barely distinguishable from man's own," theologians urged the laity to scrutinize their mental processes on a daily basis. As Johnstone observes, the clergy insisted that the "first task of the godly upon waking should be to examine their minds and flush out any thoughts that the Devil had placed there whilst they slept."[29] William Perkins in *The whole treatise of the cases of conscience* exhorts his readers to learn how to distinguish between the thoughts "that are from the Devill" and one's own thoughts. There are signs or symptoms, he argues, that people can identify to assess whether the Devil has entered their minds. The Devil's thoughts, for example, "come speedily, as lightning into a house . . . forced into the minde by violence," and they "come into the minde againe and againe, yea a thousand times in a day." Devilish thoughts, Perkins maintains, "weaken the memorie, dull the senses, wearie and confound the braine." Moreover, the Devil's thoughts are "against the very light of nature," against natural knowledge, reason, and conscience.[30] Conversely, Richard Capel in *Tentations, their nature, danger, cure* (1633) asserts that it is too "hard for most men" to determine "the difference betwixt the independency and suddennesse of Satans tentations" and our own corrupt motions.[31] Even scrupulous contemplation of the nature of one's own thoughts may not help a person know if the Devil has clouded his or her reasoning or understanding.

Thomas Heywood's characterization of demonic threats in *The hierarchie of blessed angellss*, a didactic poem on the spiritual realm published in 1635, reflects the beliefs of the period in its emphasis on the Devil's capacity to infiltrate the mind and in its exhortations to readers that they practice vigilance. When a person feels raging passions in his breast, she can be assured the "Tempter's there." The Devil's goal is "to compasse and invade," so we must watch to be sure "there be no entry made."[32] Since demons are able to "molest our sleepe," enter "Our bodies," and "distract our braine," it is necessary to exercise caution at all times.[33] The Protestant Devil was a clever opportunist, so constant attentiveness proved even more necessary under the most mundane or routine circumstances.[34] In contrast to the fantastic stories of demonic possession, demonic temptation occurs without fanfare during everyday situations, when one bickers with a spouse, for example, or gossips casually with neighbors. The Devil waits for folks to feel the slightest pangs of anger or pleasure, and then he takes advantage of their relaxed state, most often in a domestic space. As Johnstone observes, the home is where Protestants located most demonic experiences.[35]

In his domestic tragedy, *A Woman Killed with Kindness*, Thomas Heywood takes up the commonplace theme of infidelity to tell a story of a man and woman yielding to diabolic temptation. Even before Anne Frankford succumbs to Wendoll's seduction, Heywood hints at the Devil's capacity to invade not only one's domestic space but also one's thoughts in subtle, almost imperceptible ways. In an important reading of the play, Nancy Gutierrez recognizes that Heywood explicitly identifies the seducer Wendoll with the Devil. As a result of her affair with Wendoll, Gutierrez claims, Anne suffers from demonic possession. In particular, Anne's decision to abstain from eating (after she has been banished from her home) points to the controversies in the period over the Puritanical practice of implementing prayer and fasting as a non-Catholic means of exorcism.[36] Gutierrez rightly discerns demonic influences on the characters; however, she must also concede that Anne shows no physical symptoms of possession.[37] If Anne is demonically possessed, then Heywood's presentation of that possession lacks the typical sensationally visible marks. A subtler understanding of early modern diabolic threats allows us to see that Heywood stages not a demonic possession but the more common experience of demonic temptation, instigated by the Devil's manipulation of Wendoll's and Anne's cognitive processes.

Since the Protestant Devil rarely presents himself in the flesh, Heywood is faced with the challenge of theatrically representing an invisible force. In *The Witch of Edmonton* (1621), John Ford, Thomas Dekker, and William Rowley would later attempt to resolve this difficulty by depicting the Devil as a dog whom only some townsfolk can see. Heywood opts, instead, to intimate that the Devil's presence enters the Frankford home through Wendoll, whose own struggles with diabolic forces convert him from Satan's victim to a satanic tempter. Indeed, he voices aloud the moments when his thoughts surrender to diabolic provocations. By the end of the play, he functions as the Devil's instrument: his story seems to mirror the supposed fate of an unregenerate laity whose damnable actions and behavior were thought to transform them into demoniac people.[38] As a potentially godly character, Anne Frankford's experience of temptation manifests itself as an infection of her conscience. Anne allows the Devil first to cloud her thinking and then to draw her toward spiritual despair. From a reformist perspective, the discovery of Anne's adultery and the "kind" punishment her husband administers help free her from Satan's hold, giving her a deathbed opportunity to seek God's help.

Heywood begins the play with a wedding celebration of Anne and John Frankford's union, establishing the marriage as idyllic but also insinuating that temptation—like the serpent in Eden—already lurks in their home. Much of the scene is devoted to the male wedding guests' praise of Anne as "A perfect wife . . . meek and patient" (1.37).[39] Sir Charles Mountford describes her as a "chain of gold to adorn" (64) Frankford's neck, adding that the match itself is notably harmonious:

> There's equality
> In this fair combination; you are both scholars,
> Both young, both being descended nobly.
> There's music in this sympathy; it carries
> Consort and expectation of much joy. (66–70)

Frankford will later confirm his agreement with Mountford's impressions in a soliloquy, asserting that he feels content to have "a fair, a chaste and loving wife,/ Perfection all, all truth, all ornament" (4.11–12). But Wendoll, too, is in attendance at the party. Although he speaks only two lines in the scene, his first statement in the opening banter foreshadows the play's tragic turn:

> *Sir Francis:* Some music there! None lead the bride a dance?
> *Sir Charles:* Yes, would she dance "The Shaking of the Sheets,"
> But that's the dance her husband means to lead her.
> *Wendoll:* That's not the dance that every man must dance,
> According to the ballad. (1.1–5)

Sir Charles's joke intentionally misconstrues the meaning of the original ballad, lewdly alluding to the physical consummation of marriage. Out of step with Charles's light-hearted ribbing, Wendoll awkwardly reminds the guests that "The Shaking of the Sheets" is about the dance of death, a medieval genre that warned folks that death did not discriminate. While the other speakers make merry, Wendoll's statement introduces more sober thoughts on the vanity of life and fleshly pursuits. The implicit mix of eroticism and death in these opening lines also anticipates Anne's downfall, in which the "shaking of the sheets" will lead directly to her deathbed.

At Wendoll's next appearance in the Frankford household (when he arrives to deliver the news of Sir Charles's match with Sir Francis), Heywood tips off his audience that Wendoll may harbor devilish tendencies. Frankford's servant, Nicholas, a melancholic character (2.23–24), reveals in an aside that he feels an inexplicable animosity toward Wendoll:

> I do not like this fellow by no means;
> I never see him but my heart earns.
> Zounds! I could fight with him, yet know not why;
> The devil and he are all one in my eye. (4.83–86)

As a kind of clown figure who provides commentary on the action, Nicholas functions as a narrative guide for the audience, encouraging them to see Wendoll as a potential threat to the Frankfords' domestic serenity.[40] Although Wendoll's likeness to a Devil may not be visible to the play's spectators, or to the other characters, Nicholas's presentiment creates enough suspicion that contemporary viewers may have observed Wendoll with increased vigilance. In the same way that early modern folks in their day-to-day lives were urged to scrutinize their neighbor's actions as well as their own thoughts for signs of the Devil's presence, Heywood implicitly provokes theatregoers to view Wendoll with circumspection.

In stark contrast to Nicholas, Frankford feels no qualms about Wendoll's character. Indeed, Frankford finds great pleasure in his company. He sees Wendoll as a gentleman "pressed by want," but also "full of quality and fair desert" (4.30–31; 66). Acting with prodigious hospitality, Frankford offers Wendoll his table, his purse, and he advises his wife to use him "with all thy loving'st courtesy" (78). During his stay, Wendoll becomes Frankford's constant companion. As Wendoll describes it, their bond becomes a physical dependency for Frankford:

> He cannot eat without me,
> Nor laugh without me. I am to his body
> As necessary as his digestion,
> And equally do make him whole or sick. (6.40–43)

Frankford's bounty and Wendoll's indebtedness frame the tortuous conflict Wendoll feels as he contemplates betraying his host. Wendoll's inequitable friendship with Frankford ensures that his seduction of Anne entails not only lust but also treachery. Even when he reminds himself that his heart is "joined and knit together" (50) with Frankford's, Wendoll proves unable to resist his baser thoughts and desires.

Grappling with temptation, Wendoll reveals in soliloquy a vehement and terrified response to the bent of his own thoughts. He enters scene six alone and "melancholy," brooding on his "love" for Anne and denouncing himself a "villain" for "apprehend[ing]/ But such a thought" (6.1–2). He describes his soul as "drenched and drowned in red tears of blood" (7), a psychophysiological state in which the "motions of the Blood and spirits" would make him an easy prey for demonic manipulation.[41] He then attempts to pray to see "if God within my heart [will]/ Plant better thoughts" (8–9). Echoing the early modern temptation stories related by alleged victims of the Devil, Wendoll's efforts at praying repeatedly and involuntarily transmute into unwelcome and immoral thoughts. As Robert Bolton and other religious writers observe, Satan's notions tend to rush into one's mind "violently, forcibly, furiously."[42] Likewise, Wendoll experiences a "violence" in being "hurried to [his] own destruction" by some unnamed pressure (17–18). With imagery that hearkens back to the medieval vice figures, Wendoll begins to embrace his role of villain, imagining that his intention to sin will "Print in [his] face/ The most stigmatic title of villain" (84–85).

Wendoll's internal dialogue culminates in a question of whether he can think and act autonomously. He insists to himself that he will not betray Frankford, arguing "I may choose,/ And I will choose" (92–93). But his resistance appears to be weakened again by an elusive force:

> If I say
> I will not do it, what thing can enforce me?
> What can compel me? What sad destiny
> Hath such command upon my yielding thoughts?
> I will not. Ha! Some fury pricks me on;
> The swift Fates drag me at their chariot wheel
> And hurry me to mischief. Speak I must—
> Injure myself, wrong her, deceive his trust! (95–102)

The reference to a fury pricking him on connotes the ambiguous status of a material Devil who acts on his victim from without and within; Wendoll experiences these thoughts and passions as foreign, commanded by an entity other than himself. It is his lament that an indeterminate "thing"—his destiny, a fury, the Fates—enforces and compels him to yield to ideas and actions that violate everything he knows to be good. Early modern audiences would be inclined to identify this "thing" as Satan.

When Wendoll makes his desires known to Anne, her first response is that heaven would never allow Wendoll to "hatch such a disloyal thought" (110), suggesting that she instinctively recognizes the devilish nature of his thinking. Anne reminds him of the friendship and honor that Wendoll owes her husband, but Wendoll's answer reveals that his moral struggle is long over. He not only accepts the name of villain, but he also insists that he will "hazard all" (129) for Anne, implying that he risks his soul "to wear rich Pleasure's crown" (112). While Anne seems to capitulate to Wendoll's rhetorical persuasions, she primarily acquiesces to the strange confusion she feels within herself. She recognizes that her "soul's health" is dependent on keeping "precious" the love she bears her husband (139–140), but she also feels that her "soul is wand'ring and hath lost her way" (150). As Henry Wilkinson declares, it is a Christian's duty to resist the "wandring, vain, and unbelieving thoughts" that Satan will implant in us.[43] At the moment Anne submits to Wendoll, she describes herself as "enchanted," as if she had entered a maze that "will prove the labyrinth of sin" (159–160). Nicholas, who witnesses their first kiss, comments on the scene, urging the audience to see Anne's infidelity as instigated by the Devil: "It is that Satan hath corrupted her/ For she was fair and chaste" (178–179). The best explanation, in other words, for Anne's immoral behavior is the Devil's infectious influence.

Historian Frank Luttmer observes that reformist theologians argued for a perceived difference in how degenerate and regenerate Christians would react to the Devil's temptations. For the unregenerate, the Devil's temptations were experienced as a conflict between reason and will, often leading them to dismiss the Gospel as foolishness.[44] As Thomas Gataker asserts, the "wicked and prophane persons vsually desire nothing more, delight more in nothing, then in transfusing of their wickednesse and prophanenesse vnto others."[45] In contrast, when tempting the godly, the Devil would lure him or her not to carnal sins but "to despaire, perswading him that his sinnes are so horrible, his life so wicked, that God will not forgive him."[46] As Wendoll and Anne's affair develops, Heywood indicates that they experience the Devil's temptations in different ways. For Anne, her sinful acts strike her with fear and anxiety:

> Oh, what a clog unto the soul is sin!
> We pale offenders are still full of fear;
> Every suspicious eye brings danger near,
> When they whose clear hearts from offence are free,
> Despise report, base scandals to outface,
> And stand at mere defiance with disgrace. (11.103–108)

Wendoll laughs at her anguish and mockingly tells her she talks "too like a Puritan" (109). In Powell's *Spiritual experiences*, one victim of demonic temptations laments in particular that "Satan did stir up many Instruments to call me Puritane . . . [which] did make me ashamed to go in publick."[47] Anne protests that she first yielded to Wendoll's desires from "want of wit," but she now finds herself submitting to him "through fear" (112–113). She hopelessly laments that they are now so far gone in wickedness, "o'er shoes," that they may as well head "straight o'er head in sin" (114). Wendoll reacts to Anne's distress with demonic exultation: he describes his own soul as jocund and "joyful above measure" at the prospect of resuming their transgressions (115). Wendoll's delight suggests that he now fully embraces his role as the Devil's vassal.

Anne's despair deepens when her husband Frankford discovers her guilt. She declares herself unworthy of forgiveness and grace, as "Lucifer from heaven" (13.81). Still influenced by demonic thoughts, she asks to be tortured to "whip" the "scandal out," hazarding the "rich and dear redemption of [her] soul" (137–138). But for Frankford, Anne's sinfulness poses a material threat to his home environment. Indeed, when their children are brought forth, he insists on their removal for fear that Anne's "adult'rous breath" will "blast their spirits/ With her infectious thoughts" (126–127). His decision to ban her from the house may stem not only from a desire to punish her with passive "kindness" but also from a fear that her demonic corruption is contagious. As Luttmer points out, the Devil's influence was interpreted as a transmittable disease. Thomas Gataker contends, for example, that "evil company is as infectious as evill aiers are."[48]

Once exiled, Anne states her plan, that she will never again "eat, nor drink, nor taste/ Of any cates that may preserve my life" (16.101–102) nor sleep nor rest until her tears "have washed my black soul white" (104). As Gutierrez argues, Anne's decision to fast may have invoked, for some audience members, the controversy over the Puritan John Darrell's claim that prayer and fasting helped cast the Devil out of the possessed. Samuel Harsnett's initial condemnations of exorcism were directed at the notorious practices of John Darrell. But as Darrell contends, in defense of his practices, prayer and fasting also functioned as "appropriate natural medicine to a natural disease."[49] Indeed, prayer and fasting had a complicated status in early modern Protestant writings.[50] In *The holy exercise of fasting* (1604), Nicholas Bownde asserts that fasting helps to:

> turne away the wrath of the Lord from themselves, and from their houses . . . . But especially when God hath taken any of them [family] away, and so hath come nearer unto us with his correction, then had we much more cause thus in fasting to cast down ourselves at his feet in the humble confession of our sinnes which were the cause of it.[51]

As an immediate and potentially infectious danger to her family, Anne's fasting could be her attempt to prevent further damage to the Frankfords' domestic world.

While Anne's refusal to eat may help purify her physically and spiritually, her final encounter with the Devil takes place before she reaches her deathbed. Plagued by the horror of his own "guilty soul" and the "scourge of repentance" (31–32), Wendoll feels compelled to see Anne once again. His degenerate nature reemerges, however, when the sight of her prompts fire and "unchaste desire" in his heart (68–69). When Anne sees him coming toward her, she classifies the moment as Satan's final temptation:

> The devil doth come to tempt me ere I die.
> My coach! This fiend, that with an angel's face

> Courted mine honour till he sought my wrack,
> In my repentant eyes seems ugly black. (107–110)

As Kristen Poole has noted, the demons that early moderns reportedly met at their deathbed were invisible to others;[52] in Anne's case, her exceptional vision allows her to see beyond Wendoll's fair appearance and describe the demon he has become. Wendoll then runs away, in Jenkin's words, "like a sprite" (121). Although Wendoll claims to carry Anne's life and sins upon his head, his plans for the future underscore a debased nature: he intends to wander in Europe until the rumors "abate" and he can return to England as a courtier (130–134).

Following her pronounced rejection of Wendoll, Anne's final scene may have been interpreted by early modern spectators as the aftermath of a tempted woman's battle with Satan. Surrounded by friends and family, her dying hour resembles Katherin Brettergh's last moments, who requested on her deathbed that her family "bring in some godlie man that may put holy things into my mind, whereby I may be able to resist Satan."[53] Anne is reunited with Frankford, who pardons her "rash offence" and asserts that her "repentant tears" unite their souls (17.107–109). Married once again to Frankford, Anne dies on the prayer that her soul will be "in heaven . . . free" (121). Whether or not Anne is saved by God, her story would remind the audience that the godly wage battle with Satan on a daily basis. Where modern audiences see a woman trapped by patriarchal strictures, early moderns potentially saw a victim of the Devil's routine enticements. Indeed, for some audience members, Anne's final hours could be seen as triumphant. For even after suffering the infection of demonic thoughts and yielding to temptation, it appears she has defeated the Devil in the end.

Heywood's staging of diabolic temptation captures an interim period in the intersecting histories of demonology and cognition. While the Devil still appeared on the English stage as a visible entity in the seventeenth century, his invisible presence proved much more difficult to combat. As a secret poison, or infection, or dangerous thought, the Devil could darken an individual's understanding, undermining his or her capacity to resist temptation. Indeed, it was the threat of demonic corruption that challenged early modern theologians on whether to interpret despair as a state of mind, a sign of damnation, a spiritual affliction, a mental illness, or a supernatural disease.[54] The secret influence of the Devil also meant that seemingly protected and familiar spaces harbored greater dangers and risks than the recognizably foreign: for Heywood and his Protestant contemporaries, demonic intrusions in the household mimicked the process of internal temptation. Far from a safe haven, the home is the frontline for struggles with an insidious and invasive Devil, and the domestic tragedy proves the apt genre for representing this spiritual warfare.

NOTES

1. Paula McQuade, "'A Labyrinth of Sin': Marriage and Moral Capacity in Thomas Heywood's *A Woman Killed with Kindness,*" *Modern Philology* 98.2 (2000): 231–250; esp. 232–233.
2. On the Protestant devil's capacity to plant thoughts, see Darren Oldridge, *The Devil in Early Modern England* (Stroud, Gloucestershire: Sutton Publishing, 2000), 45–52. In *The Devil and Demonism in Early Modern England* (Cambridge: Cambridge University Press, 2006), Nathan Johnstone demonstrates that identifying whether the "origin of the individual's innermost thoughts . . . were to be attributed to oneself or to the devil" as a central problem in English Protestant discussions of diabolical agency (24).
3. See John D. Cox, *The Devil and the Sacred in English Drama, 1350–1642* (Cambridge: Cambridge University Press, 2000), 9. Cox identifies secularization as the goal in a number of critics' works, including David Bevington, Bernard Spivack, Robert Weimann, Keith Thomas, and Stephen Greenblatt. Also see Jan Frans van Dijkhuizen, *Devil Theatre: Demonic Possession and Exorcism in English Renaissance Drama, 1558–1642*, on the common but false critical assumption that "late sixteenth-century and early seventeenth-century English culture developed both a secular and skeptical attitude towards the demonic" (Woodbridge, UK: D. S. Brewer, 2007): 1–23; esp. 20. See too the historical study, Sarah Ferber, *Demonic Possession and Exorcism in Early Modern France* (London and New York: Routledge, 2004).
4. See Johnstone, 11.
5. Stephen Greenblatt, *Shakespearean Negotiations: The Circulations of Social Energy in Renaissance England* (Berkeley: University of California Press, 1988), 94–128.
6. Greenblatt, 109.
7. Greenblatt, 105.
8. See Kristen Poole, *Supernatural Environments in Shakespeare's England: Spaces of Demonism, Divinity, and Drama* (Cambridge: Cambridge University Press, 2011). See too Michael MacDonald, "The Fearfull Estate of Francis Spira: Narrative, Identity, and Emotion in Early Modern England," *Journal of British Studies* 31.1 (1992): 32–61. Writing a cultural history of emotion, MacDonald traces how the story of Italian Protestant Francis Spira's conversion and subsequent despair was variously interpreted as a spiritual affliction, a mental illness, and a supernatural disease; indeed, many early modern writers cite Spira's emotional distress as the effects of diabolical instigation.
9. For example, see Hugh Trevor-Roper, *The European Witch-Craze of the Sixteenth and Seventeenth Centuries* (New York: Harper & Row, 1968); Keith Thomas, *Religion and the Decline of Magic: Studies in Popular Beliefs in Sixteenth and Seventeenth Century England* (London: Weindenfeld & Nicholson, 1971); Brian P. Levack, ed., *Possession and Exorcism* (New York & London: Garland Publishing, 1992); James Sharpe, *Instruments of Darkness: Witchcraft in Early Modern England* (Philadelphia: University of Pennsylvania Press, 1996). Challenges to the rationalist perspective include Lyndal Roper, *Oedipus and the Devil: Witchcraft, Sexuality and Religion in Early Modern Europe* (London and New York: Routledge, 1994); Diane Purkiss, *The Witch in History: Early Modern and Twentieth-Century Representations* (New York and London: Routledge, 1996).
10. Michael MacDonald, ed., *Witchcraft and Hysteria in Elizabethan London: Edward Jorden and the Mary Glover Case* (New York and London: Routledge, 1991), ix.

11. On early modern engagement with hysteria, see Kaara L. Peterson, *Popular Medicine, Hysterical Disease, and Social Controversy in Shakespeare's England* (Burlington, VT: Ashgate, 2010).
12. Edward Jorden, *A briefe discourse of a disease called the suffocation of the mother* (London, 1603), B3r.
13. Stuart Clark, *Thinking with Demons: The Idea of Witchcraft in Early Modern Europe* (Oxford: Clarendon Press, 1997). See Michael MacDonald's description of Richard Napier's medical practice, for example, in *Mystical Bedlam: Madness, Anxiety, and Healing in Seventeenth-Century England* (Cambridge: Cambridge University Press, 1981).
14. Stuart Clark discusses the preceding physicians in "Demons and Disease: the Disenchantment of the Sick (1500–1700)," in *Illness and Healing Alternatives in Western Europe*, ed. Marijke Gijswijt-Hofstra, Hilary Marland, and Hans De Waardt (London and New York: Routledge, 1999): 38–58; esp. 40–42.
15. See David Harley, "Explaining Salem: Calvinist Psychology and the Diagnosis of Possession," *The American Historical Review* 101.2 (1996): 307–330.
16. Philip C. Almond, *Demonic Possession and Exorcism in Early Modern England: Contemporary Texts and Their Cultural Contexts* (Cambridge: Cambridge University Press, 2004): 24–30.
17. On exorcism as a political battlefield, see MacDonald, *Witchcraft and Hysteria*, xix–xxvi. On the community importance of possession and exorcism, see Nancy A. Gutierrez, "Fasting and Prayer in *A Woman Killed With Kindness*: Religious Salvation and Political Resistance," in *"Shall She Famish Then?" Female Food Refusal in Early Modern England* (Aldershot, UK: Ashgate, 2003): 35–51; esp. 44–45.
18. John Cotta, *The triall of vvitch-craft shewing the true and right methode of the discouery: with a confutation of erroneous wayes* (London, 1616), 53.
19. William Perkins, *The combat betvveene Christ and the Diuell displayed: or A commentarie vpon the temptations of Christ: preached in Cambridge by that reuerend and iudicious diuine* (London, 1606), 7; also quoted in Johnstone, 72–73.
20. Johnstone, 75.
21. William Gouge, *The vvhole-armor of God: or A Christians spiritual furniture, to keepe him safe from all the assaults of Satan* (London, 1619), 81, 70–71.
22. John Downame, *The Christian warfare Wherein is first generally shewed the malice, power and politike stratagems of the spirituall enemies of our saluation, Satan* (London, 1604), 25.
23. Richard Gilpin, *Demonologia sacra, or, A treatise of Satan's temptations in three parts* (London, 1677), 25.
24. A. W., *The young-mans second warning-peece* (London, 1643), 4.
25. Vavasor Powell, *Spirituall experiences* (London, 1653), 143.
26. Powell, 145, 357–368.
27. Anon., *The Christian life and death of Mistress Katherin Brettergh* (London, 1612), B1r.
28. See for example Richard Norwood's protracted battle with Satan in *The Journal of Richard Norwood* (New York: Scholars' Facsimiles and Reprints, 1945), 90–111.
29. Johnstone, 61, 115.
30. William Perkins, *The whole treatise of the cases of conscience* (Cambridge, 1606), 162–163.
31. Richard Capel, *Tentations, their nature, danger, cure* (London, 1633), 36.
32. Thomas Heywood, *The hierarchie of blessed angells* (London, 1635), 492.
33. Heywood, *Hierarchie*, 492, 497.

34. Johnstone, 155.
35. Johnstone, 76, 100.
36. Gutierrez, 43.
37. Gutierrez, 46.
38. Frank Luttmer, "Persecutors, Tempters and Vassals of the Devil: The Unregenerate in Puritan Practical Divinity," *The Journal of Ecclesiastical History* 51.1 (2000): 37–68.
39. All quotations from *A Woman Killed with Kindness* refer to the scene and line numbers in the edition in *Plays on Women*, ed. Kathleen E. McLuskie and David Bevington (Manchester: Manchester University Press, 1999) and appear parenthetically in the text.
40. For speculation on the staging of the first production of *A Woman Killed with Kindness*, see Robert Cardullo, "The First Production of Thomas Heywood's *A Woman Killed with Kindness*," *Studia Neophilologica* 83.2 (2011): 199–210.
41. See Gilpin, 25, discussed earlier.
42. Quoted in Oldridge, 46.
43. Henry Wilkinson, *Counsels and comforts for troubled consciences* (London, 1679), 25.
44. Luttmer, 57.
45. Thomas Gataker, *The Spiritual Watch* (London, 1622), 68.
46. Luttmer, 55.
47. Powell, 217.
48. Quoted in Luttmer, 62.
49. Almond, 10.
50. For a helpful discussion of fasting in the period, see Tom Webster, *Godly Clergy in Early Stuart England: The Caroline Puritan Movement, 1620–1643* (Cambridge: Cambridge University Press, 2003), 60–74.
51. Nicholas Bownde, *The holy exercise of fasting* (Cambridge, 1604), 219, 236–237.
52. Poole, 66.
53. *The Christian life and death of Mistress Katherin Brettergh*, B4r.
54. See MacDonald, "The Fearfull Estate," for a discussion of despair as etiologically confusing (50–53).

5 Mental Bodies in *Much Ado About Nothing*

James A. Knapp

> The soule indeede is so diuine, that raising and mounting itself sometimes aboue all natural formes, it comprehendeth by an admirable, absolutely free, and incompulsiue power, all incorporeall things seuered and divided from all matter and substance.
>
> —Helkiah Crooke, *Mikrokosmographia*

The justification for Claudio's infamous rejection of Hero at the altar in *Much Ado About Nothing* raises an epistemological problem that Shakespeare returned to often: the gap between perception and knowledge, particularly the knowledge of another person. Claudio bemoans "what authority and show of truth/ Can cunning sin cover itself withal!" before going on to charge that despite her "exterior shows," Hero "knows the heat of a luxurious bed;/ Her blush is guiltiness not modesty" (4.1.34–35, 39–41).[1] Claudio then draws attention to the signifying power of Hero's "seemingly" guiltless face, only to emphasize that her appearance does not reflect her inner character:

> Out on thy seeming! I will write against it:
> You seem to me as Dian in her orb,
> As chaste as is the bud ere it be blown;
> But you are more intemperate in your blood
> Than Venus, or those pamp'red animals
> That rage in savage sensuality. (4.1.56–61)

A principal irony in the scene is that Claudio's certainty—he *knows* the truth about Hero's character—is a product of his own eyes, the very evidence he rejects in the process of Hero's public shaming.[2] Even more telling perhaps is that he is willing to draw attention to Hero's ambiguous appearance in support of his claim ("her blush is guiltiness, not modesty"). Unlike Hamlet who *is* as he *seems* despite his suggestion that the two are not necessarily connected, Claudio is certain that Hero *is not* what she *seems*. Later in the scene, Friar Francis will also point to Hero's appearance, but he will do so to make his case for her innocence. Why can Claudio believe what he has seen in the dark from a distance more fervently than what his eyes tell him in the light of day before the altar? And, assuming he is looking at the same person, why can the Friar be so sure that Claudio is wrong?

The scene highlights the relationship between perception and cognition and the underlying philosophical puzzle known as the "mind-body problem." Specifically, in what manner does the sensuous perceptual body inform thought or cognition, the mental activity associated variously in the period with the physical brain, the intermediary mind, and the immaterial rational soul? In *Much Ado About Nothing*, Shakespeare invites his audience to contemplate the relationship between Hero's outward appearance and her immaterial character, a relationship that preoccupied early modern natural philosophy, medicine, and theology, resulting in complicated and often contradictory explanations. My particular interest here is in the status of the immaterial in these explanatory systems and specifically in how Shakespeare exploited contradictions in contemporary accounts of immateriality to open up theatrical possibilities in a play aptly titled *Much Ado About Nothing*. The well-known pun of the play's title—nothing as "noting"—points to the mental activity of processing the input of the senses through cognition at the same time that it references the immateriality of the thoughts produced by such activity. Before turning to the play directly, it is useful to take a brief survey of the medical and philosophical thought on the subject of body-mind interaction in the period. Having done so, I then turn to the specific question of how these theories of physical and mental activity inform an early modern understanding of perceptual recognition and misrecognition before examining the play's exploration of these themes in the final section.

BODY/SOUL, BRAIN/MIND, SPIRIT

In the first chapter of his *Mikrokosmographia* (1615), an encyclopedic tome devoted to the anatomical structure and functioning of the body, Helkiah Crooke takes a moment to contemplate the essence of the body's immaterial other: the "soule, the most excellent of all formes."[3] Noting that the subject of his book is the body, "the measure and exemplary patterne of all corporeal things," Crooke contrasts the body to the incorporeal soul, which "comprehendeth . . . all incorporeall things severed and divided from all matter and substance." Indeed, he continues, "if it could be discerned with the eye, or but conceived by the mind, how it would ravish us, and leade us to an excessive love of itself?"[4] The inaccessibility of the soul through both sense perception and mental cognition leads the proto-scientist to abandon the effort: "Let us content our selves to handle that that may be handled, or at least is subject unto some of our sences, and so proceed to the other part of Man, namely, the Bodie, which more truly and properlie is the subject of our Discourse."[5]

Although Crooke's opening digression on the essence of the soul reflects conventional wisdom of the time, his decision to include it in a text about the body and his confused rhetorical efforts to establish meaningful distinctions between body, mind, soul, spirit, thought, and so forth reflect the extent to which early modern advances in anatomy highlighted this developing science's inability to coherently articulate the relationship between mind and body. Mental activity (or the thinking brain) is particularly problematic for a writer like Crooke, one hoping to establish a meaningful distinction between the corporeal (and thus material) body and the incorporeal (and, it would follow, immaterial) soul. This passage, at least, places mental conception on the corporeal, material side of the divide along with discerning "with the eye," in keeping with the Galenic tradition underpinning Crooke's account.

While steps toward establishing the modern distinction between mental activity and the functioning of the physiological body would come only decades after Shakespeare produced his final plays, at the turn of the seventeenth century the debate was far from settled. The conventional description of the relationship between body and soul, brain and mind based on a mixture of Aristotelian, Galenic, and scholastic thought had become confused, some of its long-standing tenets called into question by the emergence of observational science and changing attitudes toward theological questions brought on by the Reformation. At the heart of the changing debates was neither the status of the body, the materiality of which was clear, nor the soul, at least initially understood to be immaterial (a requirement for eternal salvation), but an intermediary substance, called spirit. Crooke acknowledges the scope of the debate in a brief section devoted to the "nature of a spirit":

> Because in the Schooles of Physicians the Controversie concerning the naturall spirit is sufficiently bandyed, I will not spend much time in a thing so notorious: onely for their satisfaction to whom these subtleties are most strange and lesse obvious, I will give a taste or short assay concerning the nature of spirits: the distance is not so great between the highest Heaven and the lowest Earth, as is the difference betwixt the Soule and the Bodye. It was therefore verie necessarie that a spirite should bee created, by whose intermediate Nature, as it were by a strong though not indissoluble bonde the Divine soule might be tyed to the bodie of earth. . . . Our definition of a spirit shall be this, *A subtle and thinne body always mooveable, engendered of blood and vapor, and the vehicle or carriage of the Faculties of the soule.*[6]

Crooke's solution for bridging the "great distance" is to follow Galen in defining the corporeal spirit as the "thinnest of all" substances, but a material substance nonetheless.[7] For Galen, as James J. Bono explains, *pneuma* ("breath" commonly translated as "spirit") "derives chiefly from the air taken into the body (through inspiration) and from the blood . . . . It is, consequently, of a corporeal, that is material nature, though the matter in question is of an exceptionally fine and rarified sort, rather like hot vapor."[8] For Galen, concerned entirely with the body in his medical works and committed to an empirical method, the idea of "material spirits" was noncontradictory. Yet by the time Crooke was compiling the centuries of Galenic commentary that had been filtered through medieval Arabic and European scholastic thought, the relationship between the immaterial soul and the material spirit was beginning to pose more of a problem.

The idea of spirit as life-giving breath, a third term between soul and body, can be traced to antiquity, coming down to the early moderns from Aristotle and his interpreters. Aristotle's theory of *pneuma* influenced the use of the term *spirit* in commentaries on both medicine and the soul (most notably Aquinas and Avicenna), which use the term *spirit* to refer to either a fine material substance or an incorporeal body, such as an angel. As Christine Göttler has pointed out, for Aquinas *spirit* refers to a broad range of "subtle" entities: "It can be said of corporeal as well as incorporeal things."[9] His sense of the term as a designation for "subtlety" allows him to assign it to everything from

> the warming and vitalising breath, the air inhaled and exhaled by the lungs . . . that is to say in the sense of *pneuma*, a key term for Galenic medicine . . . [but] also the more noble elements of air and fire . . . [and] the corporeal virtues and spirits that move inside the body and rule and regulate the operations of the human mind.[10]

Part of the confusion over the term resulted from the odd status of spirit as aid or vehicle to bodily or intellective functions in the different theories of Galen and Aristotle. Gad Freudenthal highlights the difficulty in his explanation of Aristotle's theory of "vital heat"—that which is necessary for animate life:

> Vital heat necessarily inheres in a substrate—it is not a substance existing on its own. . . . For the action of vital heat on blood, from which all the other bodily parts are formed, necessarily produces in it *pneuma*, and it is this *pneuma* which in fact is the immediate substrate of vital heat.[11]

The flexibility of the term as a stand-in for something between the corporeal and incorporeal, material and immaterial, in medical, philosophical, and theological discourse from antiquity through the Middle Ages led ultimately to an untenable situation in the early modern period. According to Bono:

> the sixteenth century accepted the blurring of difference between nature and the divine . . . . Such tendencies had their foundations in contingent social and cultural aspirations without which the boundaries containing

> speculative Galenic theoretical medicine might have remained more narrowly defined. As it was, the broader definitions of the sixteenth century placed intolerable strains upon the language inherited from the Middle Ages and new languages of life and theology reconstituted themselves to form what is known as the scientific revolution.[12]

Later in the seventeenth century, arguments over the nature of material and immaterial substances would intensify, animating philosophical debates involving Descartes, Hobbes, Spinoza, Leibnitz, and Locke.[13] Yet during the period marked roughly by Shakespeare's earliest production during the last decades of the sixteenth century and the publication of Descartes's influential *Meditations* in 1641, interest in the relationship between mind and body was increasingly focused on defining the difference between things material and immaterial.

During this transitional period, the prevailing opinion on these matters was that the thinking mind (or brain) was in the corporeal realm, separated from the immaterial soul. An even more radical version of this view is found in Hobbes's objections to Descartes's *Meditations*, printed along with the author's responses in 1641. Hobbes's objection to the meditation on the nature of the human mind, for example, rests on the notion of the thinking mind as a corporeal substance:

> since knowledge of the proposition 'I exist' depends on knowledge of the proposition 'I think,' and the knowledge of this latter proposition depends on the fact that we cannot separate thought from the matter that thinks, it seems that we must infer that a thing that thinks is material, rather than immaterial.[14]

By the time Hobbes was arguing in favor of what was increasingly viewed as a radical form of materialism, the tide of opinion had turned toward Descartes, in favor of a notion of ideas as immaterial substances. An example suggestive of the arc of the century's debates can be found in John Harris's 1698 Robert Boyle lectures, published as *The Atheist's Objections, Against the Immaterial Nature of God and Incorporeal Substances, Refuted.* In response to "the strange Absurdities of the contrary Opinion, *That there is nothing but Matter in the World,*" Harris asserts "That the Notion of a Spirit, or an Immaterial Substance, is as Intelligible as that of Body; and that we have as much reason to believe the Existence of the former, as of the latter."[15] To support this view Harris calls on the Cartesian language of clarity and distinctness:

> And it is most certainly true that the idea which we have of Body doth not necessarily include Cogitation in it, nor our Notion of Cogitation include Body: but they are two as distinct Ideas as any we have. So far are they from being the same thing, that we cannot possibly conceive

> Cogitation with Extension. No Man ever conceived a Thought to be so many Inches or Yards long; to be deep, thick or broad, to be divisive into two or more parts, or to have any kind of figure or determinate Position or Extension.[16]

Harris's clarity is a result of decades of philosophical debate and express efforts on the part of those interested in empirical observation to disentangle religious matters from the domain of science and philosophy.[17] (Hobbes himself stated that the nature of God was beyond philosophy.) That the title page describes Harris's work as "two sermons preached at the Cathedral Church of Saint Paules" but also the "4th and 5th lectures for that year founded by the honorable Robert Boyle, esq." is suggestive of the pressure felt by members of the Royal Society to ensure its work not come under suspicion of atheism. But at the beginning of the seventeenth century—before these lines were drawn—distinctions between the material and immaterial were less clearly allocated to the realms of science and religion respectively.

The corporeal spirit described by Crooke is a peculiarly ambiguous substance, an intermediary or go-between that shares something of the body's corporeality while simultaneously able to touch the incorporeal realm and make contact with the immaterial soul.[18] Cognition is similarly associated in the period with both the affected, material body and immaterial ideas or thoughts. Before Cartesian dualism, the idea of spirit as intermediary, so thin as to elude the senses but yet material, must have seemed particularly attractive. As Justin Smith explains, Descartes's familiar dualism requires a clear distinction between the material and immaterial that was simply not relevant to the medieval tradition still mainstream in the sixteenth century. Smith quotes sixteenth-century French physician Jean Fernel on spirit as representative of the rear-guard view: "the name 'spirit' is given to any corporeal or incorporeal thing that does not enter into the senses. . . . All immaterial substances that escape the senses exercise their influence on material bodies by means of spirit."[19] Shakespeare's interest in "nothing"—no thing, the immaterial—is of particular interest at a moment when Descartes's hard distinction between the mind and body was not yet in place but the waning and confused view of liminal substances seemed increasingly untenable. It is in this context that I would like to consider the "mental bodies"—the cognitive, immaterial substances—we encounter in Shakespeare's play.

THE PERCEPTUAL BODY AND THE BODY PERCEIVED

Much Ado About Nothing repeatedly dramatizes the role of the perceived body in shaping both the characters' and the audience's ideas about the principle persons of the play. The first exchange between Beatrice and the messenger sets up the thematic interest in one of the play's key terms—*nothing*. Ignoring the messenger's report of Benedick's "good service," Beatrice comments on his character in the language of the body, calling him a "very valiant trencher-man" with an "excellent stomach" (1.1.44, 46–47). The messenger's attempt to return the conversation to Benedick's reputation, by generously interpreting her comment on his stomach as a compliment to his courage, reaches its peak in his declaration that Benedick is "stuffed with all honorable virtues" (ll.51–52). This honorific statement produces a curious restraint in Beatrice, who exclaims that "It is so indeed, he is no less than a stuffed man; but for the stuffing—well, we are all mortal" (ll.53–54). Beatrice's comment can be read as an application of Falstaff's catechism on honor in *1 Henry IV*, as Benedick's honorable virtues are reduced to stuffing, or simply stuff (*OED* "dust"), the corrupt materiality that defines our mortality (and humanity). The exchange is enough to prompt Leonato to intervene, proclaiming Beatrice's banter the result of a "merry war," the product of which, Beatrice is quick to add is "nothing" (ll.56, 59). Her subsequent quip, that in their last "conflict, four of his five wits went halting off, and now is the whole man governed with one" (ll.59–61), further adds to the play's thematic emphasis on the gap between the material and immaterial, corporeal and incorporeal. The insult relies on the distinction between the five wits and the five senses as internal and external human capacities (mental faculties and physical senses).

Beatrice's "five wits" are generally taken to be the mental faculties of "memory, imagination, judgment, fantasy, and common sense."[20] She concludes her jab with a conventional insult to Benedick's intelligence: "so that if he hath wit enough to keep himself warm, let him bear it for a difference between himself and his horse, for it is all the wealth that he hath left to be known a reasonable creature" (1.1.61–65). If one considers mental activity to be material—as the examples in Crooke's account suggest was current at the time—Beatrice's jab can be seen to draw on not only the distinction between the internal faculties and the external senses but also the interrelations among them that depend on the medical spirits.[21] In other words, although Beatrice is referring to the mental faculties here, the mental realm was not so distinct from the sensuous as it appears in more modern, post-Cartesian accounts. Daniel Hellar-Roazen has suggested that the Aristotelian account of the mental faculties, and the common sense in particular, may have been aimed at sensation as much as cognition, or rather, that consciousness—that which we associate with mental activity—was more closely connected to the sensuous than the reflective. As opposed to reflection in the modern sense of inward contemplation, Heller-Roazen proposes an "inner touch."[22] To explore the implications of this in early modern England, it is useful to turn again to medical accounts of cognition (still primarily Galenic) that were influenced by Aristotle and his commentators.

According to the French surgeon Ambroise Paré, writing at the end of the sixteenth century:

> Actions [of the body] are two-fold; for they are either Naturall, or Voluntary. They are termed Naturall, because they are performed not by our will, but by their owne accord and against our will: As are that continuall motion of the Heat, the beating of the Arteryes, the expulsion of the Excrements, and such other like which are done in us by the Law of Nature whether we will, or no.[23]

The mind only enters into the discussion when Paré turns to the issue of the "voluntary actions," which he divides further into the three subsets "sensitive, moving, and principle." The first two categories are overtly bodily—the "sensitive" associated with the organs of sense, and the "moving" resulting from the motion of muscles produced "either by bending and contraction, or by extension." He then goes on to describe what would seem to be the properly mental faculties:

> The Principall Action and prime amongst the Voluntary is absolutely divided in three, Imagination, Reasoning and Memory.
>
> Imagination is a certaine expressing, and apprehension which discernes and distinguisheth betweene the formes and shapes of things sensible, or which are knowne by the senses.
>
> Reasoning is a certaine judiciall æstimation of conceived or apprehended formes or figures, by a mutuall collating, or comparing them together.
>
> Memory is the sure store of all things, and as it were the Treasurie which the minde often unfolds and opens, the other faculties of the minde being idle and not imployed . . . . all the forementioned Actions whether they be Naturall, or Animall and voluntary, are done and performed by the help and assistance of the Spirits.[24]

According to Paré, the spirits similarly fall into three categories: the animal, vital, and natural. These, in turn, are conceived in gradations of materiality, the most refined being the animal (so called "not because it is the Life [margin: anima], but the chief and prime instrument thereof"), which is "made and laboured in the windings and foldings of the veins and Artereyes of the braine" and is so ethereal a substance that it is like to air.[25] When the animal spirit is obstructed, the person can appear lifeless "sometimes wanting both sence and motion."[26]

At the other end of the spectrum we find the natural spirit, which "hath its station in the Liver and Veins. It is more grosse and dull than the other, and inferior to them in the dignitie of the Action and the excellencie of the use." Whereas the role of the vital spirit is to "nourish the heate which resides fixed in the substance of each part [of the body]," the natural spirit serves "to helpe the concoction both of the whole body, as also of each severall part, and to carry blood and heate to them."[27] The natural spirit is at once the least substantial in terms of dignity (that which separates humans from animals) and the most grossly material (beastly as in basic bodily function).[28] Paré notes that there is some debate as to the existence of the natural spirit, perhaps due to the incompatibility of its gross materiality and the intermediary character of the spirits as conduits between the material body and immaterial soul.[29] The elegant division of labor among the animal, vital, and natural spirits allows for the complex voluntary actions to function properly.

Beatrice's insult, quite simply that Benedick is dim-witted and thus not a worthy mental sparring partner for her, carries with it the implication that his physiological-mental interface is corrupt. Bret Rothstein notes the implication of the connection between the material spirits and mental activity in late medieval medical discourse: "Mental function, though frequently considered in terms of its relationship with the soul, is bound up with the state of the body . . . . Variegation of intelligence arises from the degree to which and ease with which spirits move about within the body."[30] When the two future lovers exchange wits for the first time only lines later, their exchange follows a conversation between Leonato and Don Pedro concerning Hero and specifically Hero's identity as Leonato's daughter. This fact, they all agree, is confirmed by her appearance: "Truly the lady fathers herself" (1.1.101–102), which is intimately intertwined with her character: "Be happy, lady, for you are like an *honourable* father" (ll.102–103; emphasis added). The conversation offers a contrast to Beatrice's lowest insult to Benedick in the same scene. In response to Benedick's mocking relief that she will not seek the company of men, "so some gentleman or other shall scape a predestinate scratched face," she disparages his appearance: "Scratching could not make it worse, an 'twere such a face as yours were" (ll.123–127).

All this banter—animated by language highlighting the divide between bodily appearance and internal character—introduces the initial catalyst for the comedy's action: Claudio's apprehension of Hero as "a modest young lady": "In mine eye she is the sweetest lady that ever I looked on" (1.1.153, 174–175). This most crucial act of perception happens before we learn of it in the dialogue, as Claudio only reveals his love after Hero exits. Once alone with his friend, Claudio asks if Benedick has "noted" Hero, hoping, apparently, to confirm through her outward appearance that she is worthy of his love. Benedick's response, "I noted her not, but I did look on her" (l.153), emphasizes the difference between apprehension and comprehension—between Hero's body provided by the senses and her body as seen through the lens of cognitive judgment. In Paré's terms Benedick admits only to the act of "imagination" while Claudio's objection that Benedick "speak in sober judgment" invokes the category of "reasoning."[31] Reasoning requires the third category of actions described by Paré: memory "the sure store of all things." For in order to compare and estimate, one needs more than one term. Claudio's claim that Hero is "the sweetest lady that ever I looked on" is comparative: as far as he can judge, based on the evidence of his senses ("in mine eye"), Hero is sweeter than any other lady. Claudio's positive judgment here provides the template for the play's subsequent action: by drawing on this process of comparing the "mental bodies" of memory and imagination (bodies perceived through the senses), the play realizes its remarkable scenes of both recognition and misrecognition.

RECOGNITION AND MISRECOGNITION

Claudio's ability to recognize "modesty" in Hero's outward appearance is only the first of the play's many examples of the proper exercise of judgment—the act of reasoning that allows one to compare the objects provided to the mind by the senses (Pare's "apprehended formes or figures") to the objects of prior experience. Claudio applauds Don Pedro for just such an ability when he notes that the prince is able to "know love's grief by his complexion" (1.1.293). Such examples suggest that the mental actions of imagination and reasoning interpenetrate one another in the experience of material bodies.[32] While the imagination enables one to recognize what it is that distinguishes one sensible shape from another (Claudio from Benedick, for example, or Hero from Margaret), judgment allows one to determine what difference the difference makes. While the play sometimes suggests that the inner truth of a character is available in his or her outward appearance (written onto the body), internal character and external appearance are continually correlated ("collated" in Paré's term) through the exercise of imagination, reason, and memory.

The intertwining of perceptual and mental bodies is evident when Don John's claim to Conrad—"I cannot hide what I am" (1.3.12–13)—is confirmed at the beginning of the masked ball. In response to Beatrice's comments about Don John's appearance, "How tartly that gentleman looks!" Hero agrees that "He is of a very melancholy disposition" (2.1.3, 5). Later in the ball scene, Ursula recognizes Antonio through his costume "by the waggling of [his] head" and his "dry hand," and Borachio recognizes Claudio "by his bearing" (2.1.105, 108, 149).[33] Beatrice and Benedick likewise recognize each other and use the opportunity to trade barbs under the pretense of anonymity. In such cases, when everything is working correctly, recognition is a result of the human capacity for cognition: the act of comparing the impressions that material bodies make on the senses to the mental bodies retained by the memorial faculty, judging or estimating the extent of their similarities and differences. This process is fairly straightforward when the substance to be recognized in sensuous experience is the material form of the body in question. In extrapolating this process to the recognition of a person's "disposition," things get a bit messy.[34] The crucial point here is that the material or immaterial status of thoughts, ideas, and mental images was in question at the moment that the play was produced. Importantly, in early modern medical and philosophical accounts, material bodies are never directly compared in the act of cognition because they must first be converted into mental images by the imaginative faculty. In the case of the more abstract kinds of knowledge about human character—the kinds of observations about temperament, loyalty, and trustworthiness that Shakespeare explicitly explores in *Much Ado*—the language of the body used in the process of cognitive understanding is explicitly divorced from sense experience and corporeality but not necessarily the materiality of the person doing the thinking.

E. Ruth Harvey cites Avicenna's discussion of the weakened cognitive abilities of the drunkard to illustrate this point. Avicenna argued that "wine is too damp for serious thought: drinkers concern themselves with the pleasant rather than the profitable, the present and not the future."[35] "The reason for this," explains Avicenna:

> is that the power of the soul performs its actions in the brain, and it needs spirit of moderate humidity, which will obey the motions of cogitation, and the operations of intellect. But [after drinking] the spirit is very moist and. . . . [t]herefore it obeys the intellect and the motions of cognition only in very material and corporeal matters, not in anything very subtle or incorporeal; and, in its instability and agitation, it will not serve to form or present immaterial things, but only to form the most corporeal ones.[36]

As such explanations suggest, more complex understanding—the proper domain of the rational soul—requires contemplation of immaterial things, formed for contemplation by the cognitive faculties. The positive gain in divorcing cognition from the immediacy of sensual experience is to avoid the influence of strong sensation to inflame the passions and thus cloud judgment. The negative, as I believe Shakespeare was well aware, was to enter a realm of detached mental contemplation without any material reality against which to check the validity of one's judgment. If, as Aristotle claimed, everything begins with the senses, the movement away from sense experience in mental experience has the potential to lead thought into a world of confusion.

In the case of *Much Ado*, the confusion comes when the characters are invited to compare mental bodies derived from the imagination's ability to create false images in the mind without any immediate external stimulus. As Rothstein argues, "Imagination is the crux of the matter, for it is the faculty that provides judgment not only with things seen but also with items that result from mental fabrication."[37] Philosophers since antiquity had noted the human ability to contemplate things that cannot have been derived from the senses. The examples offered to make this philosophical point tend to be extreme: "an emerald man or a flying mountain."[38] But the observation also suggests that it is possible to contemplate mental fabrications that are false but possible in the world of sense experience. This idea is particularly important for the dramatist whose interest is in creating plausible illusions. The confusions that animate the action of *Much Ado* are precisely of this sort. Consider the gulling of Benedick. Prior to Don Pedro's manipulation in the garden, Benedick responds to Beatrice's appearance ("look here she comes") with disdain: "here's a dish I love not" (2.1.246, 257). But after he hears the description of "the effects of passion" that Hero has supposedly witnessed in Beatrice, his mind is altered. The fiction offered to Benedick's ear in the gulling scene reconstructs his memory of Beatrice's body—her mental body in his cognitive experience—by endowing it with the "effects of passion" vividly described by the plotters: "she tore the letter into a thousand halfpence; railed at herself, that she should be so immodest to write to one that she knew would flout her. . . . Then down upon her knees she falls, weeps, sobs, beats her heart, tears her hair" (2.3.138–140, 144–145). The very next time he sees Beatrice approach, his judgment of her appearance is entirely different: "Here comes Beatrice. By this day, she's a fair lady! I do spy some marks of love in her" (2.3.239–241).

At this point in the play, Benedick's belief that he sees "some marks of love in" Beatrice is an example of misrecognition, or misestimation (the failure of properly judging the data of the senses). In fact, Benedick contemplates an imaginative fiction: the "love-struck Beatrice" is plausible but not actual, as no such material form exists in the play. Having yet to be gulled by Hero and Ursula, Beatrice would not yet bear the signs Benedick sees; they are apparently the product of Benedick's comparison of the material body standing before him on stage and the immaterial, mental body lodged only moments before in his mind by Don Pedro, Claudio, and Leontato. This harmless tutoring of Benedick's judgment sets up the much more sinister scene of misrecognition with which I began this essay—the misrecognition of Margaret as Hero that leads to Hero's public shaming. According to Borachio, the misrecognition of Margaret for Hero on the balcony is the product of three things: Claudio and Don Pedro are fooled "partly by [Don John's] oaths, which first possessed them, partly by the dark night, which did deceive them, but chiefly by my villainy, which did confirm any slander that Don John had made" (3.3.150–154). Considering that Borachio is bragging of his villainy to Conrad, his claim that he played the chief role is dubious at best. In fact, although each of the factors is important for the deception to come off, it is Don John's framing of the events to come that determines the perceptual experience for both Claudio and Don Pedro. The dark of the night and Borachio's theatrical performance (calling Margaret by Hero's name) ensures that neither Claudio nor Don Pedro rely on the evidence of their eyes in a simple comparison of Hero's bodily appearance to the one before them on the balcony. They misrecognize Margaret as Hero in large part because the object of comparison, Don John's planted story—"you shall see her chamber window entered, even the night before her wedding day" (3.2.101–103)—gives them something to compare with the scene they witness on the balcony. This something is in fact nothing, but this revelation simultaneously confirms how important nothing—the immaterial—is in the process of human judgment. Don John's admonition, "If you dare not trust that you see, confess not that you know" (3.2.113–114), places special emphasis on the evidence of the senses—and particularly vision—while at the same time deflecting scrutiny of his fabricated story of Hero's infidelity. Don John's comment to Claudio that "it would better serve your honor to change your mind" (ll.104–105) than marry Hero doubles as the catalyst to literally change Claudio's mind. Importantly this is not the scene that Shakespeare chooses to stage.

Rather than show us the scene in which Margaret's outward appearance is mistaken for Hero's, Shakespeare chooses to stage the much more jarring scene in which Claudio tragically misrecognizes Hero's inner character at the altar. In this scene, the playwright emphasizes how Hero's outward appearance, and in particular her blush, can simultaneously signify the truth of her inner character and the misguided view of the deceived Claudio and Don Pedro. The blush is a particularly important marker of inward character in the period, as Levinius Lemnius confirms in his *Touchstone of Complexions* (1576):

> There is no surer way (sayeth Galene) certainly to knowe the humours and juyce in a Creature, then by the colour and outward complexion. If the body loke very white, it is a token that phlegme in that body, chiefly reigneth and most aboundeth. If it be pale or yellow, it argueth the humour to be greatly Melancholique and Cholerique, and the blood to be fresh and reddye: if it be blackish, it betokeneth blacke . . . Choler.[39]

According to this kind of humoral observation, one's disposition is available on one's face. However, Lemnius goes on to assert that both "outwarde accidental occasion" and internal emotional states can alter the appearance of the complexion. The complexion is a good guide

> if no outwarde accidentall occasion happen, as great heat or chaffing, labor or wearinesse: or if the mynde bee not intoxicate, and perplexed with affects and passions, as Anger, Joye, Sorrow, Care, pensiveness: for these make the humours sometime to resort unto the skin and utter parts, and sometime to hyde and conveyghe themselves farre inwardly: and for this cause we see men that are . . . testy to be in a marvelous heat, proceeding not from any sickness or disease, but of the motion and stirring of the humours: againe, them that be affrighted and in mynde amazed, to be pale.[40]

Thus, despite humoral theory's taxonomic potential for deciphering disposition, identifying inward character from the face is a difficult matter in practice. Claudio admits that Hero's blush can betray either modesty or guilt depending, he suggests, on the condition of her conscience. The more sinister revelation of the scene, however, is that the signifying power of the blush is actually reliant on the state of mind of the beholder. I have already commented on Claudio's response to the blush—as evidence of her guilt. What is particularly striking about the scene is how Friar Francis draws on the same visual evidence to come to the opposite conclusion:

> I have marked
> A thousand blushing apparitions
> To start into her face, a thousand innocent shames
> In angel whiteness beat away those blushes,
> And in her eye there hath appeared a fire
> To burn the errors that these princes hold
> Against her maiden truth. (4.1.158–164)

The Friar attributes his ability to discern Hero's "innocent shames" from guilty blushes to mental training: his "reading" and his "observations. . . which with experimental seal doth warrant the tenor of [his] book" (4.1.165–167). His is a more discerning eye, owing to a training in recognizing appearances confirmed through years of experience in comparing perceptual and mental bodies. His subsequent plan relies on this ability, as he proposes to reconstruct Hero's mental body in Claudio's brain:

> When he shall hear she died upon his words,
> Th'idea of her life shall sweetly creep
> Into his study of imagination,
> And every lovely organ of her life
> Shall come appareled in a more precious habit,
> More moving-delicate, and full of life,
> Into the eye and prospect of his soul,
> Than when she lived indeed. (4.1.223–230)

The mental body that the Friar describes exceeds the one still present on the stage.[41] The innocence that Hero's material body is unable to confirm, her mental body will. It is the mental body, "more moving-delicate, and full of life" that has the power to touch the "prospect of his soul." The "thousand blushing apparitions" recall the animating, thin spirits that bring life to the human body. Despite the living Hero's presence on the stage during this scene, in hypothetically recuperating Hero's reputation for the absent Claudio, the Friar uses the language of bodily spirits to bring her back to life.

This is all the more striking in the context of the theatre, where Hero's blush is most likely to be known to the audience through the dialog rather than her ruddy face. As Michael Neill has recently pointed out, most spectators would not have been close enough to the actors to read facial expressions, much less recognize an actor's blush. But the detail is crucial for the action, and Shakespeare clearly exploits the commonplace that the blush conveyed an ambiguous significance.[42] What Don John has done to Claudio and Don Pedro and what Claudio has in turn done to Leonato is precisely what Shakespeare is doing to the audience throughout the play. This intertwining of the material body ("every organ of her life") with the mental body and ultimately the soul is foregrounded in Benedick's response to the Friar's plan: "I will deal in this/ As secretly and justly as your soul/ should with your body" (4.1.247–249). Benedick's reassurance that he will "deal justly" is underwritten by his characterization of the soul's relationship to the body—the soul deals with the body both "justly" and "secretly." The inaccessibility of the soul, and its resulting secrecy, highlights the distinction between body and soul, material and immaterial that the play exploits. Later, when he is challenged by Beatrice to judge the truth of Claudio's accusation, Benedick blurs the line between body and soul, asking "Think you in your soul the Count Claudio hath wronged Hero?" (4.1.327–328). In asking this, Benedick hopes for certainty. But Beatrice's response, "Yea, as sure as I have a thought or a soul" (4.1.328–330), reinscribes the problem that is at the center of the play, thoughts and souls cannot be known materially, and as such they elude the kind of certainty Benedick seeks. Beatrice's response gestures at the gap between the total comprehension of the soul and the thinking mind, dependent on material substances for its subject matter. One can think only with the embodied mind, and, even if the soul can know absolutely as Crooke suggests, the two are always divided by an unbridgeable gulf.

Although successfully rescuing Hero's reputation from slander, the Friar's intervention in defense of proper judgment fails to resolve the more troubling implications of the mysterious interplay between the material body, corporeal spirit (thought), and immaterial soul that Shakespeare explores throughout the play. The various examples of recognition and misrecognition in *Much Ado* reveal both the fallibility of the senses and the fallibility of reason. If, following my epitaph from Crooke, "The soule . . . is so diuine, that . . . it comprehendeth . . . all incorporeall things seuered and divided from all matter and substance," this is precisely what the embodied mind cannot do. Benedick notes the dependence of thought on the body: "if their wisdoms be misled in this,/ The practice of it lives in John the Bastard,/ Whose spirits toil in frame of villainies" (4.1.187–189). The practice of misleading is intertwined with Don John's body—"frame"—wherein "spirits toil."[43] Similarly, in rejecting Antonio's advice to use reason against his grief for Hero's defamation, Leonato specifically points to the inability of the intellect to counsel one who is impassioned bodily:

> men
> Can counsel and speak comfort to that grief
> Which they themselves do not feel; but tasting it,
> Their counsel turns to passion, which before

> Would give preceptial medicine to rage,
> Fetter strong madness in a silken thread,
> Charm ache with air, and agony with words. (5.1.20–26)

Leonato's insistence that he be allowed to feel his grief—"I will be flesh and blood" (5.1.34)—is set against things that border the immaterial: "preceptial medicine," "air," and "words." The "silken thread" capable of fettering "strong madness" is especially suggestive of spirits as the thinnest of substances.

The thoughts that slander and redeem in this play are embodied in the minds of Shakespeare's characters. These mental bodies, being dependent on rather than free from matter, are subject to its distortions, to the vagaries of sense experience, to suggestion, and to time. Leonato's entrance upon learning of the plot to defame Hero encapsulates the problem:

> Which is the villain? Let me see his eyes,
> That when I note another man like him,
> I may avoid him. Which of these is he? (5.1.253–255)

Although his desire to know villainy by its outward appearance is understandable, the play offers little hope that it can be accomplished. Claudio's attempt at contrition muddles the question of what even counts as villainy: "Impose me to what penance your invention/ Can lay upon my sin; yet sinn'd I not/ But in mistaking" (5.1.267–269). This weak defense of his actions highlights the deeper problem of acting on what one takes to be immaterial truths—another's virtue, love, or ill will, for example. Without access to the soul's unfettered comprehension of things immaterial, cognition in the embodied mind is messy, unpredictable, and ultimately bound to the flesh of material bodies. As it is represented here and elsewhere in Shakespeare, even the disciplined mind offers no guarantee that we will recognize innocence or guilt by its outward appearance. Just ask Othello.

NOTES

1. Unless otherwise indicated, quotations follow the Arden Shakespeare, *Much Ado About Nothing*, ed. A. R. Humphreys (New York: Methuen, 1981; Rpt. Thompson Learning, 2004). Not surprisingly, this scene has attracted a great deal of critical attention. See, for example, Carol Thomas Neely, *Broken Nuptials in Shakespeare's Plays* (New Haven, CT: Yale University Press, 1985); Carol Cook, "'The Sign and Semblance of Her Honor': Reading Gender Difference in *Much Ado About Nothing*," *PMLA* 101 (1986): 186–202; Nova Myhill, "Spectatorship in/of *Much Ado About Nothing*," *Studies in English Literature* 39 (1999): 291–311; Andrew Fleck, "The Ambivalent Blush: Figural and Structural Metonymy, Modesty, and *Much Ado About Nothing*" *ANQ* 19.1 (2006): 16–23.

2. For a recent study of visual perception and knowledge in the period see Stuart Clark, *Vanities of the Eye: Vision in Early Modern European Culture*, (Oxford: Oxford University Press, 2007).
3. Helkiah Crooke, *Microcosmographia* (London, 1616), Sig. B2r.
4. Crooke, Sig, B2r–B2v.
5. Crooke, Sig. B2v.
6. Crooke, Sig. Q3r–Q3v.
7. Crooke, Sig. Q3v.
8. James J. Bono, "Medical Spirits and the Medieval Language of Life," *Traditio* 40 (1984): 91–130; esp. 92.
9. Christine Göttler, "Preface: Vapours and Veils: The Edge of the Unseen," in *Spirits Unseen: The Representation of Subtle Bodies in Early Modern European Culture*, ed. Christine Göttler and Wolfgang Neuber (Leiden and Boston: Brill: 2008): xv–xxvii; esp. xx–xxi.
10. Göttler, xxi.
11. Gad Freudenthal, *Aristotle's Theory of Material Substance: Heat and Pneuma, Form and Soul* (Oxford: Clarendon Press, 1995), 130.
12. Bono, 101–102. See also 93–98 for a discussion of the medieval developments in the tradition.
13. See, for example, John Sutton, *Philosophy and Memory Traces: Descartes to Connectionism* (Cambridge: Cambridge University Press, 1998), and Justin E. H. Smith, "Spirit as Intermediary in Post-Cartesian Natural Philosophy," in Göttler and Neuber, ed., 269–291.
14. René Descartes, *Meditations, Objections, and Replies*, ed. Roger Ariew and Donald A. Cress (New York: Hacket, 2006), 102.
15. John Harris, *The Atheist's Objections, Against the Immaterial Nature of God, and Incorporeal Substances, refuted* (London, 1698), Sig. D3v, Sig. D3r.
16. Harris, Sig. E3v–E4r.
17. Smith points out that it took quite some time to abandon the idea of a material spirit (288–289).
18. See Bono; Smith; and Göttler. For the history of "spirit" as influenced by Aristotle's theory of *pneuma* or "vital heat," see Freudenthal; and Edwin Hartman, *Substance, Body, and Soul: Aristotelian Investigations* (Princeton: Princeton University Press, 1977).
19. Jean Fernel, *On the Hidden Causes of Things: Forms, Souls, and Occult Diseases in Renaissance Medicine.* With an edition and translation of Fernel's *De abditis rerum causis*, ed. J. M. Forrester (Leiden: 2005), II 7, quoted in Smith, 269.
20. Norton, 530. Arden identifies them as "wit, imagination, fantasy, judgment, and memory" (153). See E. Ruth Harvey, *The Inward Wits: Psychological in the Middle Ages and the Renaissance* (London: Warburg Institute, 1975).
21. In *Twelfth Night*, Feste (impersonating Sir Topas) urges the afflicted Malvolio to "Endeavor thyself to sleep, and leave thy vain bibble-babble" in order to aid the "heavens" to restore his wits (4.3.95–97). Similarly, Kent's concern that Lear's "wits begin t'unsettle" (3.4.160) as a result of his exposure to the storm demonstrates Shakespeare's emphasis on the connection between the inward wits and the external perceptual body.
22. Daniel Heller-Roazen, *The Inner Touch: Archaeology of a Sensation* (New York: Zone Books, 2009), 40–41 and passim.
23. Ambroise Paré, *The workes of that famous chirurgion Ambrose Parey translated out of Latine and compared with the French. by Th: Johnson* (London, 1634), Sig. C6v. Paré first published his *Works* in Paris in 1575.
24. Paré, Sig. D1r.

25. Paré, Sig. D1v. According to Paré, the animal spirit is made "of an exquisite subtile portion of the vitall brought thither [to the brain] by the *Carotide Arteria*, or sleepy Arteryes, and sometimes also of the pure aire, or sweete vapour drawne in by the Nose in breathing." This is clearly the Galenic *pneuma*, though bearing signs of commerce with Aristotelian thought.
26. Paré, Sig. D1v.
27. Paré, Sig. D1v.
28. *OED* notes that "concoction" was both associated with "digestion" and with the process of "'ripening' of morbific matter, fitting it for elimination from the living body."
29. Bono notes that the natural spirit is not found in Galen's extant texts, but was attributed to him by his followers (92). Crooke's comment that there is much debate about the natural spirit (above) refers to the fact that its presence was in doubt.
30. Bret Rothstein, "Description, Intelligence, and the Excitation of Sight," in Göttler and Neuber, ed., 47–70; esp. 58–59.
31. As noted above, Paré distinguishes between "imagination" as "apprehension which discernes and distinguisheth betweene the formes and shapes of things sensible, or which are knowne by the senses" and "reasoning" is described by Paré as "a certaine judiciall æstimation of conceived or apprehended formes or figures, by a mutuall collating, or comparing them together."
32. Sibylle Baumbach, *Shakespeare and the Art of Physiognomy* (Tirril, Penrith: Humanities Ebooks, LLP, 2008), has traced the importance of this kind of reading of outward appearances in relation to the pseudo-science of physiognomy. She quotes Sir Thomas Browne's *Religio Medici*: "there are mystically in our faces certain Characters which carry in them the motto of our Souls, in which he that cannot read ABC may read our natures" (Browne quoted in Baumbach, 24).
33. For a catalogue of the ways of reading physiognomic features and for a specific discussion of masking and physiognomy see Baumbach, chapter 4, and 162–168.
34. *OED* indicates that the early modern senses of disposition referred to both "mental constitution or temperament; turn of mind" and "physical constitution" or "physical aptitude, tendency, or inclination (to something or to do something)." ("disposition" s.v. 6., 9[a]. 9b.).
35. Harvey, 26.
36. Avicenna, quoted in Harvey, 26.
37. Rothstein, 55.
38. The examples are from Avicenna via Rothstein, 55.
39. Levinius Lemnius, *Touchstone of Complexions*, Trans. Thomas Newton (London, 1576), M2r.
40. Lemnius, M2r.
41. It is important that Hero has not left the stage through the entire scene in which her appearance is taken to signify the extremes of innocence and guilt.
42. See Fleck; and Schoenfeldt, 6–7.
43. Frame is usually glossed as "contriving" or "plotting" but also carries the meaning of bodily makeup, as in *OED*: 9. a. "Applied to the animal, esp. the human body, with reference to its make, build, or constitution."

[...Second Link...]
The Unbearable Permeability of Bodies and Minds

Michael Schoenfeldt

One of the most fascinating aspects of early modern thought is its frequent recourse to corporeal narratives to explain phenomena that we would treat as intellectual, emotional, cognitive, or spiritual. This is in part a product of anxiety; it is an effort to bestow a material cause, and so perhaps to produce the groundwork for a material solution, to matters that seem to belie human intervention. Purportedly inexplicable phenomena could be explained by recourse to the flexible fictions of humoral excess or astral influence or demonic invasion. But instead of alleviating anxiety, these fictions only produced a deeper anxiety about the necessary and inevitable permeability of the mind, lodged in a body subject to various environmental influences. By rendering cognitive and spiritual issues in largely material terms, then, early modern writers were forced to confront just how minds interact with the world through the body, and how potentially treacherous that interaction can be.

In a popular work that Shakespeare almost certainly knew, *The Book Named the Governor* (1531), in the section entitled "Examples of Continence," Sir Thomas Elyot tells a stunning story demonstrating the full terror of such external contamination:

> Valerian, being Emperor of Rome, and persecuting the Church, in Egypt a Christian man was presented unto him, whom he beholding to be young and lusty, thinking therefore to remove him from the faith, rather by venereal motions, than by sharpness of torments, caused him to be laid in a bed within a fair garden, having about him all flowers of sweet odour and most delectable savours and perfumes. And then caused a fair tender young woman to be laid by him all naked, who ceased not sweetly and lovingly to embrace and kiss him, showing to him all pleasant devices, to the intent to provoke him to do fornication. There lacked little that the young man

> was not vanquished, and that the flesh yielded not to the service of Venus; that perceiving the young man, which was armed with grace, and seeing none other refuge, he with his teeth did gnaw of his own tongue, wherein he suffered such incredible pain, that therewith the furious burning of voluptuous appetite was utterly extinct. In this notable act, I wot not which is to be most commended, either his invincible courage in resisting so much against nature, or his wisdom in subduing the less pain with the more.[1]

Here the young Christian turns toward his own body the violence that Spenser's Guyon in Book 2 of *The Faerie Queene* would turn on the Bower of Bliss, the garden of sensual delights that aroused him. Whereas Guyon destroys the Bower, Elyot's devout Christian chews off his own tongue. Remarkably, Elyot praises as an example of continence a kind of self-mutilation that silences the stirring of desire, if only by creating an even more excruciating pain. Extreme neurological distraction offers at least temporary immunity from the desires of the flesh.

The early modern period was troubled by the permeability of our bodies and by the fact that this permeability renders our minds vulnerable to external influences. But since life entails a necessary osmosis with the world through breathing, eating, and excretion, the only impermeable body is a dead body. The senses too become the sources of particular anxiety, since they function as gates that offer openings to a world of torrential stimulation and toxic contamination. This is the reason that some of the leading philosophical schools of the time promote fantasies of stoic tranquility and ascetic discipline.

It is perhaps no accident that the essays which hold the central position in this volume dedicated to the relations between body and mind focus on *The Winter's Tale*. In the famously unprovoked sudden-onset jealousy that erupts out of nowhere, physiology is cognition, body is mind. The tragic eruption of course begins with an act of self-diagnosis: "I have tremor cordis on me," Leontes announces, experiencing a physiological symptom which can be either positive or negative (that is why he must add "My heart dances,/ But not for joy, not joy" (1.2.112–113). Darryl Chalk calls appropriate attention to the critical role of vision in this moment of emotional and cognitive misfire. It is something he sees, and misinterprets, that causes Leontes's heart to race tremblingly. Vision here is the medium of contagion, as if even the most mundane visual interaction with the world is morally and physically hazardous. Body and mind could be infected merely through the apprehension of visual external phenomena. As Thomas Wright remarks in *The Passions of the Minde*: "All senses no doubt are the first gates wherby passe and repasse all messages sent to passions: but yet the scriptures in particular wonderfully exhort, command, and admonish us to attend unto the custodies and vigilance over our eyes."[2] Vision demands particular vigilance. This anxiety about vision must have produced an interesting dynamic on the Renaissance stage, a space and medium premised on the concept of plausible illusion.

After the tragic consequences of this violent visual misprision, Leontes is left to explore in sorrow the philosophical fantasy of ascetic retirement. He succumbs to the temptation of absolute self-enclosure; or, as the figure of Time aptly describes Leontes's withdrawal from the world, "he shuts up himself" (4.1.19). This action, to borrow a term that Hardin Aasand applies deftly to *The Winter's Tale*, produces the ultimate aspiration to singularity; it is perhaps the only way to avoid most of the vectors of contagion that threaten the self. But it is not an infinitely sustainable state. Other chapters likewise focus on a kind of visual anxiety, linked in particular to the complex interpretability of visual phenomena such as the blush. If what Leontes misreads is a truly innocent gesture of spousal accommodation, the blush enters far more intrinsically ambiguous territory. A blush certainly is a corporeal, and largely involuntary, manifestation of an inner state, but its meanings are hard to parse, since it can represent absolutely opposite states: it can be either a sign of innocent modesty or a mark of knowing guiltiness. By attending to the interpretive complexity of the blush, we see how the visual data received by the body-mind nexus must be filtered through the circuits of perception and cognition. Studies of *Coriolanus* and *Much Ado About Nothing* in this volume show how these plays rehearse pivotal components of *Hamlet* and *Othello*, plays in which cognitive error and emotional suspicion tragically invade the spaces of perception.

Mary Floyd-Wilson also demonstrates in this volume how so many of these anxieties are wrapped up in the imagination of the supernatural, and particularly in the existence of a Devil who has insidious power over us. The Devil's power here is twofold, and both powers have to do with an ability to enter our inmost selves: first, he has knowledge of our deepest desires, which allows him to amplify our worst tendencies and desires and dampen our best; and second, he can exercise actual material intervention over our daily lives. Regularly amplifying our passions, the Devil becomes a kind of external correlative for our darkest impulses. And in the play that Floyd-Wilson focuses on, *A Woman Killed with Kindness*, it is fascinating to watch demonic influence invade the domestic sphere, as it would later do at length in Milton's *Paradise Lost*. The Devil's reach is so extensive throughout early modern drama that it might be seen to form what Aasand identifies as a singularity—the Devil, that is, transcends Time but is inexorably linked to all that happens in time.

Fears about the pollution of the mind frequently cluster around issues of eating, since eating is an activity absolutely necessary for life. And eating, the process of inviting something into the self, is at once absolutely necessary for life and continually threatening to its integrity. Fasting is in part an effort to free the self, at least temporarily, from its dependence on the outside world for sustenance. It is perhaps for this reason that fasting was imagined to have power to repulse the Devil. Anne's fast in this play, then, is not just spiritually purgative but an effort to assert a fragile autonomy in a world that regularly denies such. Wendoll, moreover, appropriately describes his relationship to Frankford in strikingly nutritive terms:

> I am to his body
> As necessary as his digestion,
> And equally do make him whole or sick. (6.40–43)

These lines not only remind us of the precarious nature of eating but also underscore the fact that in this period (and now), mind is not brain. Rather, mind is embedded not just in the matter of the body but also in various environmental and social networks. In many ways, the plays become studies in the way these networks are disrupted and subsequently restored or damaged irretrievably.

Together, these fascinating essays greatly advance our interpretations of individual works by looking closely at the network of connections between body and mind that they repeatedly posit. These essays also further our sense of the richness and complexity of the body-mind nexus in the early modern period. That is, they demonstrate just how difficult, and how necessary, it is for us to get our own post-Cartesian minds around the pre-Cartesian predominance of material explanations for largely spiritual or psychological phenomena. We need to remember, moreover, that this emphasis on the implicit materiality of mind, for all of the harm caused by the zany theories and pernicious therapies it fostered, was at least an unwitting coconspirator in the outburst of brilliant dramaturgy on the Renaissance stage. Only a world which imagined that bodies, and bodily experience, are fully suffused with mind and rife with morality could produce a question like Banquo's after he and Macbeth met the witches: "have we eaten on the insane root/ That takes the reason prisoner?" (1.3.87). The sense that an external event can work on its viewer like a hallucinogenic herb, imprison reason, and so change behavior may have been the source of some of the most imagistically powerful and psychologically insightful poetry to traverse any stage. In the effort to discover material correlatives for immaterial sensations, then, early modern writers and thinkers transformed the occasion of immense corporeal anxiety into the matter of breathtaking lyric intensity.

[. . .]

NOTES

1. Sir Thomas Elyot, *The Book Named The Governor*, ed. S. E. Lehmberg (New York: Everyman's Library, 1962), 205.
2. Thomas Wright, *The Passions of the Minde in Generall* (1604), ed. Thomas O. Sloane (Champaign: University of Illinois Press, 1971), 150.

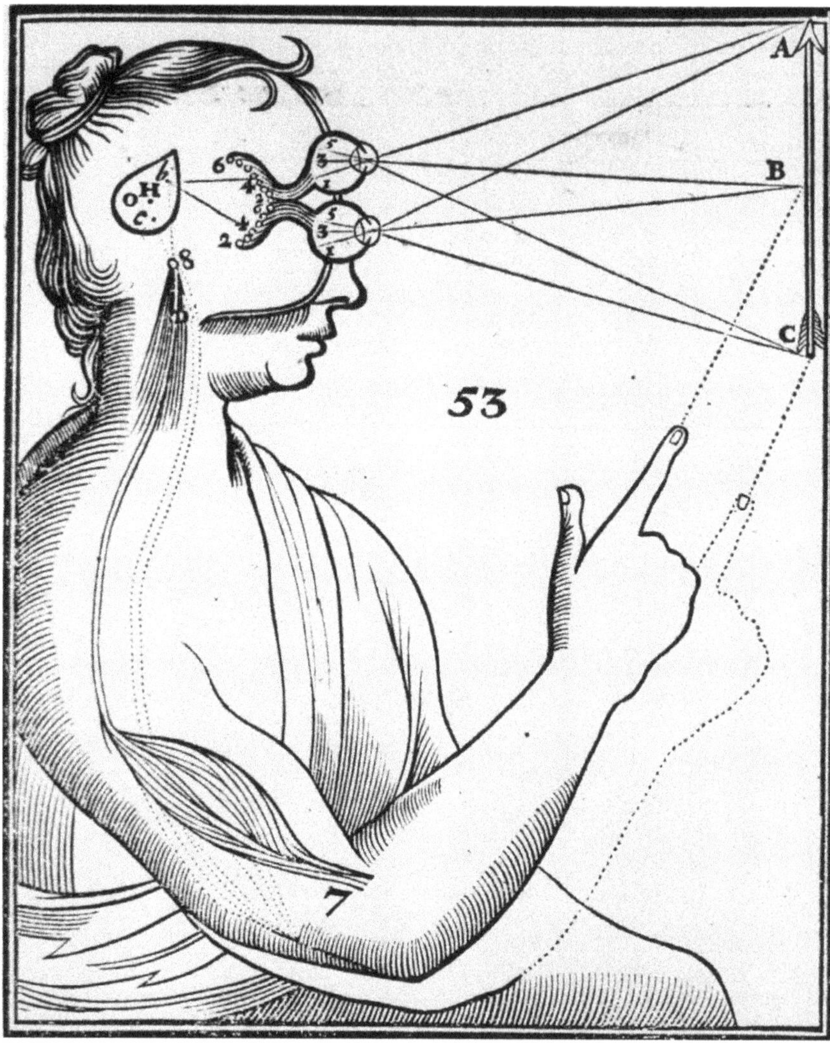

Figure 3 "Vision and the mechanism for response to external stimuli," from René Descartes, *Tractatus de homine, et de formatione foetus* . . . (1662), 139. Image A016385, History of Medicine Collection, Courtesy U.S. National Library of Medicine.

6 "Make Me Not Sighted Like the Basilisk"

Vision and Contagion in *The Winter's Tale*

Darryl Chalk

A VISION SO APPARENT

In Act One, Scene Two of *The Winter's Tale* (1611), Leontes famously undergoes a sudden transformation. As he watches his wife, the very pregnant Hermione, attempt to convince his friend Polixenes to stay longer in Sicilia, he suffers a strange alteration that leaves him in a jealous reverie, an affliction that confuses those around him. It is a remarkable change: seconds before Leontes had been calm, polite, the model of courtly behavior, and then, in an instant, observing the seemingly innocent exchange between Hermione and his childhood friend, he is struck with a rage that does not subside until his entire family has been destroyed. This precipitous shift has long troubled critics of this play, who have often found its suddenness inconceivable and dramatically implausible. John Dover Wilson, in his 1931 edition of the play, instigated the notion that Leontes's alteration was so untenable that he must have been jealous before the events of the play commence and that he only attempts to persuade Polixenes to stay as a means of testing long-held suspicions.[1] Roger J. Trienens extended this idea by suggesting that Leontes's "secret jealousy" is deliberately held as a "surprise," disclosed to the audience as a moment of revelation "characteristic of Shakespeare's genius."[2] J. L. Styan attests that Leontes's jealousy "descending like a bolt from hell" has proven an "embarrassment" to many an actor taking on this role in the play's production history and that it is a moment "we may never fully understand."[3] More searching and fruitful examinations of this issue have Michael D. Bristol viewing it as "a type of spatiotemporal derangement"; David Hillman as a "moment of corporeal skepticism"; Valerie Traub as an extreme example of anxious masculinity wherein Leontes reacts to "Hermione's pregnant body [as] a palpable reminder of erotic activity"; and Gail Kern Paster as a kind of competitive rivalry in which the maternal body triggers in Leontes a fear of usurpation by his "twinn'd lamb" Polixenes.[4] I propose, however, to re-look at this scene as an incident where the visual is principally at stake, and to suggest that if examined in the context of contemporary medical knowledge, the passionate paroxysm at its center is far from beyond our comprehension.

As Leontes himself explains, it is the perceived image of infidelity, "a vision so apparent" (1.2.272), that he sees as the cause of his radical change, and his extreme, transformative state is a disease, self-identified as "the infection of [his] brains" (1.2.147).[5] While vision has been a predominant feature of much criticism concerned with the play's extraordinary final scene, little consideration has been given to Leontes's transformation as a moment of seeing.[6] His affliction is, after all, largely related to ideas about cognition and perception in the play, and both its cause and effect can be seen as bound up with the pervasive connections between vision and contagion in early modern culture.

Indeed, the kind of infection of the mind Leontes contracts in *The Winter's Tale* can be understood in relation to theories of cognition and the passions at this time that construct the body-mind as vulnerable to pathogenic infiltration by images through its key sensory facility: the eyes. Donald Beecher has defined this notion of contagion by sight:

> Contagion is a phenomenon of transmission, generally of pathogens that attack the physical organism. *Tangere*, or touch, is the root of the word, suggesting a necessary proximity so that potential parasites may leap from host to host. Nevertheless, it was recognised throughout the second millennium of Western medicine that ideas and images, communicated from afar, retain their power to injure the mind and redound on the body with full psychosomatic potency. Thus it followed that the mechanisms of infection might be extended to include the transmission of those images that were capable of provoking states of disease in the body.[7]

This chapter suggests that *The Winter's Tale* stages an example of contagion by image in its opening scenes. It demonstrates that a Jacobean audience might have understood Leontes's abrupt change and subsequent delirium of fury as being consistent with early modern medical conceptions of diseases of the body-mind, in particular those that were thought to be caused by pathogenic communication at a distance. In such formations Leontes's affliction is shown to be a disease of the imagination, one brought about by an offensive image that infects his mind. Produced by a cognitive contagion, the condition can also be spread from mind to mind; as can be seen, I argue, in the perhaps equally perplexing—and arguably no less sudden—death of Leontes's son Mamillius. Thus two of the most peculiar issues in a particularly troubling play are not only resolvable but are intrinsically linked.

The language in the play explicitly renders Leontes's transformed demeanor as changes happening in his body and mind and emphasizes vision as a critical factor in this process. During the moment in question, he appears quite conscious of the sudden change at the exact time it comes upon him, describing (to himself or in direct aside to the audience) his own reaction to the interaction of Hermione and Polixenes as an onstage spectator looking at a scene both he and the audience can see:

> Too hot. Too hot:
> To mingle friendship farre is mingling bloods
> I have *tremor cordis* on me. My heart dances,
> But not for joy, not joy.
> [. . .] to be paddling palms and pinching fingers,
> as now they are, and making practised smiles
> As in a looking-glass; and then to sigh, as 'twere
> The mort o' th' deer—O that is entertainment
> My bosom likes not, nor my brows. (1.2.110–121)

The self-consciousness of this and other speeches, how he seems to instantaneously diagnose his own affliction, is extraordinary but also highlights the role that vision has to play in this moment. Apart from Leontes's privileged description, there are no stage directions or textual clues to tell us exactly what Hermione and Polixenes are doing at this moment; thus we have no idea how accurate his account of their body language is. Each production of this play answers the issue because it can, of course, make greater or lesser degrees of difference between their actions and the accusatory substance of Leontes's speech; but there is no way of knowing what a playgoer like Simon Forman saw in this moment at one of its initial performances at the Globe. All Forman says about this scene in his diary entry after seeing *The Winter's Tale* in 1611 is: "Observe there how Leontes, the king of Sicilia, was overcome with jealousy of his wife with the king of Bohemia, his friend, that came to see him."[8] Clearly Leontes reacts strongly to something *he sees*, or at least perceives, and it is the significance of this perception and what it does to his mind and body that I am interested in here.[9]

We gain further insight into Leontes's embodied transformation when other characters start to notice that he appears perturbed. Hermione suggests that "He something seems unsettled" (1.2.148), asking him, "You look/ as if you held a brow of much distraction./ Are you moved, my lord?" (1.2.150–152). Later, Polixenes will remark that "the king has on him such a countenance/ As he had lost some province, and region/ loved as he loves himself" (1.2.369–671); Leontes's refusal to meet his gaze ("wafting his eyes to the contrary" as Polixenes puts it) leaves him wondering "what is breeding that changes thus his manners" (1.2.375–376). Leontes himself provides perhaps the most revealing commentary on this change in a speech that has been named among the most incomprehensible in Shakespeare[10]:

> Affection, thy intention stabs the centre.
> Thou dost make possible things not so held,
> Communicat'st with dreams—how can this be?—
> With what's unreal thou coactive art,

> And fellow'st nothing. Then 'tis very credent
> Thou mayest co-join with something, and thou dost
> [. . .]
> And that to the infection of my brains
> And hard'ning of my brows. (1.2.140–149)

One of the more troublesome aspects of this passage is the word *affection*. Frequently glossed as "lust" or "passion," Hallet Smith has shown that it is closer to the Latin *affectio*, which is suggestive of a perturbation of the body-mind. He cites the definition from the 1582 edition of Cooper's *Thesaurus Linguae Romanae et Brittannicae*, which clearly describes it as such: "Affectio, Verbale. Cic. Affection: a disposition or mutation happening to bodie or minde: trouble of minde. Impetus, commotio, affectioque animi."[11] David Ward has also astutely reasoned that not only is the Folio's rendition of this speech ("Affection? thy Intention stabs the Center") more helpful than most editorial renderings of it but that it might not have been so difficult for an early modern audience to understand if we can see it in early modern terms. Ward suggests that if read in relation to contemporary conceptions of the passions, the passage makes quite a bit more sense. In this contextual reading *affection*, a word related specifically to appetite, explains and constitutes Leontes's "infection" as a kind of psychosomatic disease.[12] The word *intention* also gains medical resonance as *intensification*, a term "frequently used to describe the sudden onset and growth of unbalanced or diseased physical or mental states."[13] I seek to take up Smith's suggestion that Leontes describes a "sudden seizure" of the body and mind, extend Ward's reading of this key passage, and look in greater depth at just how an early modern audience might have understood not just this speech but Leontes's physiological and emotional well-being in these opening scenes, his precipitous shift, and its effect on others as a kind of contagion of the mind. By what and how is he "moved"? What constitutes the "*tremor cordis*" he says he feels? Can we diagnose "the infection of [his] brains" in the way that an early modern audience might have done? In what ways were such conditions thought to have the capacity to operate by contagion?

THE EYE OF THE MIND

Vision and cognition were intimately entwined in early modern culture. Of the five senses, sight was considered the principal and most reliable means of receiving information into the body. The French royal physician Andre Du Laurens, in *A Discourse of the Preseruation of the Sight*, translated into English and published in London in 1599, viewed sight as the "most noble" and "excellent" of the senses singling out "the certaintie of the function. For it is out of all doubt that this is the most infallible sense, and that which least deceiueth: according to that which men are wont to say, when they will assuredly auouch any thing, namely, that they see it with their own eyes."[14] Medical literature in sixteenth- and seventeenth-century England largely subscribed to Aristotelian theories of perception and cognition. Vision was thought to occur because, as Stuart Clark has summarized it, "objects in the world gave off resemblances or replicas of themselves (*species*) which then travelled to the eyes and, via the eyes and the optic nerves, into the various ventricles of the brain to be evaluated and processed."[15] The early moderns thought "of perception as a visual process," Clark explains: "What eventually found its way into the memory . . . was a mental picture or 'phantasm' . . . something that could be 'seen' by the 'eye' of mind."[16] Despite the apparent faith in sight by writers like Du Laurens as the most transparent and objective of cognitive processes, vision was far from fail-safe. The image could be deceptive, as in a conjuring trick or other types of legerdemain, or the individual could misperceive what they saw, and in such moments the mental picture preserved in the imagination, the part of the brain where images briefly sat before lodging in the long-term memory, would be defective: "phantasm" becoming "phantasie."[17] In fact, precisely because so much assurance was invested in the visual, instances of failure presented potentially more catastrophic circumstances for the individual, particularly to the psycho-humoral condition of the body-mind.[18] An infected imagination was seen simultaneously as among the causes and symptoms of perhaps the defining individual malady of the period: melancholia, designated by Clark as "a disease specifically of the imagination."[19]

While certain individuals were thought to be predisposed to particular humoral imbalances and thus disease was often considered an almost entirely internal phenomenon, the body and mind could also be infected and altered through the conduits by which they interacted with the world. The senses were particularly vulnerable in this process, and chief among these for impact on the passions were the eyes. Thomas Wright, in his oft-cited *The Passions of the Minde in Generall* (1604), in a chapter entitled "How senses mooue Passions, and specially our sight," calls for the individual to be vigilant in monitoring what they look at:

> All senses no doubt are the first gates whereby passe and repasse all messages sent to the passions . . . [but] of all senses, sight [is] the surest and certainest of his object and sensation . . . no sense hath such varietie of objects to feed and delight it, as this . . . no sense sooner mooueth, than this; . . . nor no sense peruerteth more perilous than this; for if the guide be corrupted, the followers will hardly escape uninfected.[20]

Sight, as the "surest" sense, is also the most susceptible to perverting the images that make their way into the requisite parts of the brain and body, "the followers," and thus capable of causing infection.

Timothy Bright's *A Treatise of Melancholy* (1586) examines in some detail the ways in which the body processes the information it receives from objects it encounters through sight and the kinds of emotional disturbance that can result. While perturbations can be caused by the humors themselves, the source can also be external: "First occasion riseth from outward things, wherein wee either take pleasure, or wherewith wee are offended: this obiect is carried to the internall senses from the outward."[21] Whether the object is found to be "pleasant" or "offensive," Bright suggests, is a determining factor in how the body will react, a response he situates as largely residing in two key organs, the brain and the heart:

> Of all parts of the body, in each perturbation two are chiefly affected: first the braine, that both apprehendeth the offensiue or pleasant obiect, and judgeth of the same in like sort, and communicateth it with the heart, which is the second part affected: these being troubled carry with them all the rest of the parts into a simpathy, they of all the rest being in respect of affection of most importance. The humours then to worke these effectes, which approach nigh to naturall perturbations, grounded vpon iust occasion of necessitie, alter eyther brayne or heart: if brayne be altered, and the obiect not rightly apprehended, then it is deliuered otherwise then it standeth in nature, and so the heart moued to a disorderly passion.[22]

Offensive objects produce a reaction from the brain and then the heart, which in turn stimulates the "humours," altering both organs, leading to misapprehensions and "disorderly passions." If the brain and heart become "ouercharged of humour, the apprehension and affection both are corrupted, and misse of their right action, and so all things mistaken, ingender that confused spirit, & those stormes of outrageous loue, hatred, hope or feare, wherewith bodies so passionate are here and there, tossed with disquiet."[23] It is by such logic that Bright argues humors like choler do not *cause* anger in and of themselves—those prone to an abundance of choler are brought to anger by external occasion. Responding to the idea that choler is caused by blood boiling about the heart, Bright contends that the heart

> is by outward causes mooued" and thus "kindled with anger, then the Bloud riseth, which being cholerick increaseth the heate, but addeth nothing to the passion . . . whereby it gathereth great heat within, which breathing out againe with reuenge, causeth through vehemencie, and suddenesse of the motion, that boyling of heat, procured of anger.[24]

Once the disordered passion has taken hold of the individual's body and mind, it can bring about extreme emotional states and work to distort the reportage of the sensory system. This is particularly the case with melancholy, which:

> for the most part is setled in the splene, and with the vapours annoyeth the heart, and passing vp to the braine, counterfetteth terrible obiects to the fantasie, and polluting both the substance, and spirits of the braine, causeth it, without externall occasion, to forge monstrous fictions, and terrible to the conceit, which the judgement taking as they are presented by the disordered instrument, deliuer over to the heart, which hath no iudgement of discretion in it selfe, but giuing credit to the mistaken report of the brain, breaketh out into that inordinate passion, against reason.[25]

In extreme cases the melancholic person can see "phantasticall apparitions," "counterfeit goblins," and things simply not there.[26] At the very least, melancholia "dispossess[es]" the sufferer from "the right vse of reason" and often "times doth present the fancie with feare, and as a man transported with passion is vtterly bereft of advisement, causeth the senses both outward and inward preposterously to conceiue" leaving them bereft with "perturbations . . . hard to be mitigated, by counsel or perswasion."[27]

With such a framework in mind, we can begin to understand the process by which Leontes is infected with the distemper that provokes his sudden alteration. What Leontes sees, or thinks he sees, is at issue here, as he is at pains to point out to Camillo, incredulous that others have not perceived what seems to him so plainly in view:

> Ha' not you seen, Camillo—
> But that's past doubt; you have, or your eye-glass
> Is thicker than a cuckold's horn [. . .]
> For, to a vision so apparent, rumour
> Cannot be mute—or thought—for cogitation
> Resides not in that man that does not think—
> My wife is slippery? (1.2.269–275)

This "vision so apparent," much like the "ocular proof" demanded of Iago by Othello, is irrefutable. Leontes's reasoning is consistent with theories of the senses that placed the eye as the most reliable means of gathering knowledge. Anyone who disagrees, who literally cannot "see" what is "past doubt," has no cognitive ability, no "cogitation," no "eyes, nor ears, nor thought" (1.2.277). As Bright suggests, however, his fixation is a perturbation, a preposterous conceit, beyond all mitigation and reason. Even though nothing at all is going on between Polixenes and Hermione—Camillo simply calls it a "diseased opinion" from which he needs to be "cured" (1.2.298–299)—the image of his childhood friend and expectant wife in close and intimate conversation, holding hands, twiddling fingers, smiling and gazing at one other is for whatever reason an "offensive object," in Bright's terminology, to Leontes. It is the "resemblances" or "*species*" given off by this image that enters Leontes's body through his eyes, afflicts first his brain, which in turn causes an almost instant reaction in the heart, which "dances," the palpitations ("*tremor cordis*") Leontes feels, as dark humours begin to congeal around his heart, increasing heat ("too hot, too hot"), as his choler rises by the occasion of what he sees, resulting in the very intensity of "affection" that "stabs the centre" (his heart) and causes the infection of mind of which Leontes appears to be so conscious in his most inscrutable speech. Clearly, despite his almost paradoxical capacity to self-diagnose the symptoms of his condition, Leontes has not "rightly apprehended" the "object" he sees. The infection of his imagination brings with it common symptoms of melancholia, as he starts to build a "fantasie" from his "Affection": "Thou dost make possible things not so held,/ Communicat'st with dreams . . . And with what's unreal thou coactive art" (1.2.140–142). Like the melancholic in Bright's work, Leontes's sensory perception becomes distorted, he begins "preposterously to conceive," losing "right use of reason," "forging monstrous fictions," will not listen to reasonable counsel and cannot be persuaded from his delusions that Hermione and Polixenes have "mingled bloods" and conceived a child. Later, Paulina will accuse him of producing no "more accusation/ Than your own weak-hinged fancy" (2.3.118–119) and that his "jealousies" are "Fancies too weak for boys" (3.2.178–179). *Fancy* was a word associated at the time with "fantasy" and with the "illusion of the senses."[28] Committing the offensive image to his long-term memory, Leontes starts to see more than was really there, as he rants at Camillo:

> Is whispering nothing?
> Is leaning cheek to cheek? Is meeting noses?
> Kissing with inside lip? Stopping the career
> Of laughter with a sigh?—a note infallible
> Of breaking honesty. Horsing foot on foot?
> Skulking in corners? Wishing clocks more swift,
> Hours minutes, noon midnight? And all eyes
> Blind with the pin and web but theirs, theirs only,
> That would unseen be wicked? Is this nothing? (1.2.287–295)

Recalling the encounter earlier in the scene, the overfriendly exchange between Polixenes and Hermione that so disturbed him then has become something more intimate than even he initially described it: noses and cheeks touch, kisses open-mouthed, laughter stopped with sighs (sighing was particularly associated with lovesickness), and then morphs into a full-blown-affair as his mind leaps to images of sexualized foot-play and stolen, secretive moments in "corners."

A SICKNESS CAUGHT OF YOU THAT ARE WELL

The *source* of his affliction thus defined, what of the nature of his disease? And why, perhaps, does he perceive this image to be so offensive? The most obvious answer to both questions, of course, is jealousy, and, as I show in this section, Leontes's suspicion is bound up in this play and in early modern treatises with the medical understanding of love and jealousy as kinds of infection, diseases of the body-mind received through the eye in an instant by contagion. Eyes were understood as dangerous conduits, simultaneously conceived not just as passive receptacles susceptible to infiltration by external material and light-borne forces but equally capable of emitting potent beams with the power to dazzle, seduce, infect, or even kill the object of their gaze.[29] An exchange of vision between individuals thus provoked the possibility of spreading disease; lovesickness and even plague were thought to be able to spread in this way.[30] This conception of sight also brought with it ancient tales about basilisks and cockatrices, mystical beasts able to kill with a glance, the evil eye, and the myth of Medusa. Levinus Lemnius's influential *The Secret Miracles of Nature*, first translated into English in 1658 but available in its original Latin edition from 1559, attempted to catalogue the "many hidden and secret things in nature, of a hidden and unknown effect." As illustration Lemnius provides a list of instances of eye-borne phenomena that despite having "an unknown" origin "shewed the cause of the effect . . . very likely, if not apparently and clearly":

> For example: The Basilisk doth kill a man by his sight . . . So the sight of a woolf, if he can but come near a man doth cause hoarseness . . . So the Feminine sex having their Monthly terms flowing from them, do make dusk the brightness of Ivory, and a looking-glasse; doth blunt the edge of a Sword, doth choak the Corn; . . . and she doth not onely deform every one that she meets, but her own self with spots and blemishes. By the same reason the eyes being corrupted with blearey'dnesse, and . . . rednesse in the eyes, or bloud-shot, doth happen to corrupt the eyes of others.[31]

These instances of magical thinking about the potent contagious effects of basilisks, wolves, menstruating women, and "bleary" eyes occur by exchange of vision. The same ideas are repeated in Francis Bacon's *Sylva Sylvarum* (1627) and in George Hakewill's *The Vanitie of the Eie* (1608), published just a few years before *The Winter's Tale* had its initial performances.[32] Hakewill's staunchly anti-Catholic text, written for a gentlewoman friend "bereaued of her sight," is a diatribe against all aspects of the visual sense, charging it "to be an immediat instrument, not only of wantonnes, but of gluttony, couetousnes, theft, idolatry, ielosie, pride, contempt, curiosity, envy, witchcraft, & in a manner of the whole rebellion, & apostasie, as well of the bodie, as the minde."[33] The litany of sins and emotional disorders detailed in Hakewill's pamphlet disparaged as stemming from vision is indicative of the power and suspicion associated with sight in post-Reformation England.[34] It is the conceptions of the eye in relation to love and jealousy, however, that provide an illuminating context to the logic of contagion in the opening scenes of *The Winter's Tale*.

Lovesickness, or love melancholia, thought to be a potentially fatal illness in medieval and early modern culture, is often figured as a communicable disease, one caused by pathogenic rays transmitted between eyes. Beecher has situated this notion of "eroto-contagion" along with other treatments of lovesickness, like those of Jacques Ferrand (*De la maladie d'amour ou melancholie erotique*, 1610) and Robert Burton (*The Anatomy of Melancholy*, 1621), in an extensive tradition of thought from Plato, Avicenna, and Galen through to its encapsulation in Marsilino Ficino's *Commentary on Plato's Symposium on Love* (1484), a text that became "an *idée force* among medical philosophers in the sixteenth and early seventeenth centuries."[35] Ficino conceived of lovesickness as an eye-borne contagion as venomous as the plague, and Beecher suggests that this medical tradition maintained a stance that:

> eyebeams have the capacity to alter blood and cause infections in the same way that stinking air . . . causes contagion . . . [and] that love is not only like a disease, but by dint of its symptoms a veritable disease, which by extension must be contracted through the transfer of pathogens from body to body—pathogens that in turn produce results similar to parasitic agents in an unwilling host.[36]

While the notion that love is generated through the eyes is a commonplace in Shakespearean drama, the association between lovesickness and eye-borne contagion is made explicit several times in his plays, most notably in *Twelfth Night*, where Olivia speaks of falling in love with Cesario, a page who is in fact a young woman, Viola, in disguise:

> Even so quickly may one catch the plague?
> Methinks I feel this youth's perfections
> With an invisible and subtle stealth
> To creep in at mine eyes. (1.5.261–268)[37]

Olivia's characterization of Cesario's "perfections" (clearly he is a "pleasant" object in Bright's formulation) and love's "plague" as an "invisible" entity that "creeps" into the body through the eyes by "subtle stealth" is strikingly similar to Biron's explanation of the epidemic of love melancholia that besets his companions in *Love's Labour's Lost*:

> Write "Lord have mercy on us" on those three.
> They are infected, in their hearts it lies.
> They have the plague, and caught it of your eyes. (5.2.419–421)

Although the connections are not made quite so explicit in *The Winter's Tale*, Leontes's transformation and the imagined affair between Polixenes and Hermione are not only constructed as infections but are also rhetorically related to notions and images of eye contagion in the play. As Leontes's fantasy expands, and he resorts to childish retorts and name-calling—"You lie, you lie,/ I say you lie, Camillo, and I hate thee,/ Pronounce thee a gross lout, a mindless slave" (1.2.301–303)—his rumination shifts from his own illness to Hermione's. When he is first left alone on stage, his anger fuels a deeply misogynistic diatribe in which he eventually paints her supposed infidelity as symptomatic of a pandemic of venery: "Physic for it there's none. It is a bawdy planet . . . Many thousands on's/ Have the disease and feel't not" (1.2.201–208). He later suggests that Hermione has been contaminated by Polixenes:

LEONTES: Were my wife's liver
Infected as her life, she would not live
The running of one glass.
CAMILLO: Who does infect her?
LEONTES: Why, he that wears her like a medal, hanging
About his neck, Bohemia, (1.2.306–310)

The invoking of the liver here equates Leontes's perception of Hermione's condition with the location of lovesickness in the body, as Ferrand, in the English translation of his *A Treatise on Lovesickness* (1640), suggests: "Once love deceives the eyes, which are the true spies and gatekeepers of the soul, she slips through the passageways, travelling imperceptibly by way of the veins to the liver where she suddenly imprints an ardent desire for that object that is either truly loveable or appears so."[38] Reduced to a lovesick token, a miniature portrait in a locket adorning her lover's neck, Hermione is "infected" by eyes that Leontes now murderously wishes closed as a cure for his wife's, and his own, ailment, importuning Camillo to "bespice a cup/ To give mine enemy a lasting wink,/ Which draught to me were cordial" (1.2.318–320). Later, perplexed by Leontes's increasingly erratic and evasive behavior, Polixenes demands to know from Camillo just what "is breeding/ That changes thus his manners" (1.2.375–376), and he wonders what he may have had to do with it:

POLIXENES: Your changed complexions are to me a mirror
Which shows me mine changed too; for I must be
A party in this alteration, finding
Myself thus altered with it.
CAMILLO: There is a sickness
Which puts some of us in distemper, but
I cannot name th' disease, and it is caught
Of you that are well.
POLIXENES: How caught of me?
Make me not sighted like the basilisk.
I have looked on thousands who have sped the better
By my regard, but killed none so. (1.2.381–390)

Polixenes hints here at a contagion of alteration that can be seen to spread in the first half of this play, but he also invokes the basilisk, a creature whose fatal, poisonous stare was metaphorically associated with various kinds of eye contagion throughout the period. In pressing for an answer to this puzzle of, as he says, "A sickness caught of me, and yet I well?" Camillo relates the source of the contagion not to any fatal gaze emitted by Polixenes but, again, returns to what Leontes saw: "He thinks, nay, with all confidence he swears/ As he had seen't . . . that you have touched his queen/ Forbiddenly." (1.2.414–416). Polixenes's metaphorization of himself as a basilisk is, however, still pertinent in this sense, because looking at a basilisk was thought to be capable of bringing death as readily as being trapped in its gaze. While Linda Woodbridge has suggested that normally "In the case of the fabulous basilisk or cockatrice, it is the basilisk looking at the victim while the victim is looking at the basilisk that kills, and this complex reciprocal gaze seems closest to the situation in love poetry of the Petrarchan tradition," she also contends that simply looking upon a basilisk or cockatrice was enough to infect and kill.[39] Woodbridge gives the example from *Twelfth Night*,

> When Sir Toby laughs at the challenges issued to Sir Andrew and Viola, his jest turns partly on the ambiguity of whether cockatrices kill by looking or being looked upon: "This will so fright them both that they will kill one another by the look, like cockatrices" (TN III.iv.213–215)—since both are more frightened than frightening, they will perhaps die by looking upon rather than kill by looking.

Woodbridge understands Polixenes's reference in this way, "well himself though making others sick, [he] is a sort of magical Typhoid Mary."[40] The poisonous *species* that emanate from the image of the unwitting Polixenes-as-basilisk are what infect Leontes and this is related to the connections between jealousy and lovesickness as both eye-borne contagions.

Generally, lovesickness in Shakespeare is associated with courtship rituals, and his comedies largely show the effects of love before marriage. The play does not show us Leontes and Hermione's courtship, of course, we hear only from Polixenes about the potential moment when Hermione "crossed the eyes" of his "young playfellow" (1.2.81–82). And, just before her supposed betrayal, we get a glimpse of their betrothal when we hear Leontes remind Hermione of the moment their bond was sealed, hinting at some lengthy torment before the hand-fasting could take place:

> Why, that was when
> Three crabbéd months had soured themselves to death
> Ere I could make thee open thy white hand
> And clap thyself my love. Then didst thou utter,
> "I am yours forever." (1.2.103–107)

The Winter's Tale is one of Shakespeare's few examinations of love's intensity after marriage, and jealousy was particularly associated with the marriage bond.

Treatments of jealousy as a medical condition at this time also situate it as a distemper born of the potency of love as a passion of the mind. Burton certainly viewed jealousy as such, including it in his extensive examination of lovesickness in *The Anatomy of Melancholy* and identifying it as a subspecies of love melancholia, suggesting that

> It ought to be treated of as a species apart, being . . . so furious a passion, and almost of as great extent as love itself. . . . For these causes I will [treat it] as a bastard branch or kind of love-melancholy, which, as heroical love goeth commonly before marriage, doth usually follow, torture, and crucify in like sort, deserves therefore to be rectified alike.[41]

Benedetto Varchi in *The Blazon of Jealousie* (translated into English in 1615) calls this affliction an "uncurable plague, and deadly Poyson," "which worketh most violently . . . with force and sudden power" and attests to its pathological capacity:

> When Iealousie once siezeth . . . it is pitiful to see, how cruelly it tormenteth, & how insultingly it tyrranizeth ouer them, for of all the mindes diseases, that is it, . . . such a furious perturbation, and moody Agitation it is, which throweth them into Extremities, altogether contrary to the cause.[42]

Both writers see jealousy as the almost inevitable result of excessive love, which among all of the mind's perturbations Varchi calls the most "potent and strong."[43] And if love is an infection that is received by way of the eye, the flip side of that emotional coin is jealousy as an eye-borne infection, as is shown in the anonymous pamphlet from 1593 entitled *Tell-Trothes New-yeares Gift,* which included an appended section called "Robin Good-fellowe his Inuectiue against Ielosy" which states:

> For if Jelosy be loues brother, it is by corruption of nature, brought foorth vnlawfully, which may thus be manifested. After the eye hath chosen an obiect which brings so sweet contentment to the hart . . . the eie beeing pleased with an eye, and the hart contented with a hart, they frolique both in glory as long as they rest in constancie, but wandering from forth that sanctuary, the eie either spies another eie that better pleaseth it . . . or else the eie looks currishly into his owne hart and spies some fault in himselfe which displeasing, begetteth Jelosy, whereby the eie may be said to be originall and father of both.[44]

As in Burton and Varchi, jealousy is a product of love and the eye begets both by first "choosing an object" and then exchanging it for another but also, in a phrasing that seems to resonate with the troubling self-awareness of Leontes's description of his symptoms, the "eie looks currishly into his owne heart," the jealousy born of self-doubt or "fault in himselfe."

In the decade or so leading up to the writing and first performance of *The Winter's Tale*, playgoers would have seen another example of a character self-diagnosing jealousy as a kind of infection in Ben Jonson's *Every Man In His Humour* (1598).[45] Kitely, a London merchant who fears his wife is cuckolding him with the "wild associates" (2.1.62) of his brother-in-law Wellbred, promises vigilance in watching over her: "Yea, every look or glance mine eye ejects/ shall check occasion" (2.1.199–200). After complaining of a headache and in response to his wife's mistaken diagnosis that he is suffering from "this new disease" (2.1.215), Kitely provides an astonishingly detailed account of his sudden ailment that perfectly encapsulates the theories about emotional contagion proffered in medical treatises at this time:

> A new disease? I know not, new or old,
> But it may well be called poor mortal's plague:
> For, like a pestilence, it doth infect
> The houses of the brain. First it begins
> Solely to work upon the fantasy,
> Filling her seat with such pestiferous air,
> As soon corrupts the judgement; and from thence
> Sends like contagion to the memory,
> Still each to other giving the infection.
> Which, as a subtle vapour, spreads itself
> Confusedly, through every sensitive part,
> Til not a thought or motion in the mind,
> Be free from the black poison of suspect.
> Ah, but what misery is it, to know this? (2.1.224–237)

The theory that Shakespeare himself played Kitely in the play's original production is tantalizing enough, but Jonson here confirms the relationship between vision, the imagination, and infection of the body-mind seen in writings on the passions and in *The Winter's Tale*.[46] The speech clearly figures jealousy as a plague-like disease, working its way into the body from without, and neatly articulates the stages of infection: starting in the "fantasy," the imagination which is contaminated through the eyes, the "contagion" moves through "the houses of the brain," corrupting "the judgement" and pestering "the memory" until the mind is overwhelmed with "the black poison of suspect." Tellingly, Kitely's self-awareness of the illness overtaking his mind is no help, no inoculation against infection. It is precisely such an understanding of jealousy-as-infection that Leontes proposes in *The Winter's Tale*:

> There may be in the cup
> A spider steeped, and one may drink, depart,
> And yet partake no venom, for his knowledge

> Is not infected; but if one present
> Th' abhorred ingredient to his eye, make known
> How he hath drunk, he cracks his gorge, his sides,
> With violent hefts. I have drunk and seen the spider. (2.1.41–47)

Leontes's jealousy is fundamentally connected to notions of visual perception, and he re-emphasizes in this speech the importance of ocular "proof" to this process. He infers that many might be the victims of infidelity and not know it because they have not "seen" the "spider steeped" inside "the cup," whereas the "abhorred ingredient," the intimate moment witnessed between his wife and friend, has been "present[ed]" to Leontes's "eye" so only he "has drunk and seen the spider." His distorted logic shores up his delusions; his "just censure" is an immutable and "true opinion" (2.1.39) and thus "All's true that is mistrusted" (2.1.50). The trope of poison and the idea that his "knowledge" has been "infected" by what he has seen again foregrounds the connections between contagion and eyesight that pervade this play.

Leontes's jealousy as a disease of the mind can therefore be seen as emerging from the same kind of ocular contagion as the lovesickness that might have infected him before the events of the play during his courtship with Hermione. It relies on a moment of seeing wherein the infection of the imagination is renewed and altered, shifting Leontes's affliction from the erotic melancholy of courtship to the possessive, bitter, torturous emotion that embodies love after marriage. "This jealousy," Burton avers, "which belongs to married men, in respect of their own wives. . . . A most violent passion it is where it taketh place, an unspeakable torment, a hellish torture, an infernal plague."[47] Just as the pleasant object of Hermione once "crossed his eyes" and brought with it the powerful passionate distemper of love, the offensive object of an intimate (albeit innocent) coupling of Polixenes-Hermione generates an image or object of fascination, one that infects at a distance, producing *species* or "vapours," in Beecher's understanding, that are received into Leontes's eyes, infecting his brain and heart with the contagion of jealousy, generating a sudden transformative alteration in his embodied mind with an almost magical efficacy.

IMPRESSIONS FROM MIND TO MIND

What happens to Leontes in *The Winter's Tale* thus demonstrates the early modern conception of jealousy as a disease of the imagination, a redoubtable passion of the mind, brought about by the infection of vision—but emotional maladies could also produce contagious effects, and this is abundantly apparent in the first three Acts of *The Winter's Tale.* If the impact of jealousy, as seen in Varchi and Burton, was likened to the effects of the plague, so too was one of its key symptoms, a seemingly unstoppable rage.

Anger was not only thought to be a disease caused by internal perturbations but a contagion capable of spreading from mind to mind. In a treatise specifically about anger, *Spiritual physicke to cure the diseases of the soule, arising from superfluitie of choller* (1600), John Downame is explicit about the contagious capacity of the choleric disorder: "You see then the manifold euils which waite vppon this raging disease of the minde, vniust anger, for it is not only a deadly impostume breaking out in our selues, but also an infectious & contagious plague which destroieth whole peoples and commonwealthes."[48] Leontes's contagion of "unjust anger" wreaks havoc and destructive chaos in the play's first half, manifested in the plot to poison Polixenes and his command to have the child killed, and his "violent carriage" (3.1.18) is seen to infect others in the play. When Leontes reflects upon the impact his actions have caused, he suggests that Camillo had been "transported by my jealousies/ To bloody thoughts and to revenge" (3.2.156–157) but was thankfully "tardied" in this task by his "good mind" (3.2.160). Antigonus, increasingly convinced that Leontes is "abused" and, unthinkably, Hermione "honour-flawed," threatens, that if this proves to be so, he will "geld" his three prepubescent daughters to prevent them from committing similar sins and producing "false generations" (2.1.143–150). To think of Leontes's jealousy as a contagion further highlights recurring images of disease and infection in the play: others have been "transported" by Leontes's disturbed imagination not only to "thoughts" of violence but to actual illness and death.

Polixenes and Hermione are both figured as victims of Leontes's contaminating affection in this sense. Polixenes, though "sighted like the basilisk" and already a potential source of infection, fears that Leontes's accusations, which have "altered" him, may cause even greater pollution:

> O, then my best blood turn
> To an infected jelly, . . .
> Turn then my freshest reputation to
> A savour that may strike the dullest nostril
> Where I arrive, and my approach be shunned,
> Nay hated, too, worse than the great'st infection
> That e'er was heard or read. (1.2.417–424)

Said just before he flees to safety (like those who frequently left cities during epidemics in search of less noxious air), Polixenes associates his transformed "reputation," his "best blood" becoming "infected jelly," with the kinds of olfactory, miasmic vapors thought to cause the plague, explicitly referenced here as "the great'st infection," in early modern medical pamphlets. Hermione is also linked to notions of contagion. In addition to the idea of her being infected by Polixenes, Leontes is "glad" that "she did not nurse" Mamillius, for fear that he may have been contaminated by her breast milk, but "yet" she still "has too much blood in him" (1.2.58–60).

The image of an infectious Hermione is extended to her imprisonment, a kind of quarantine where she is to be kept from human contact—"He who shall speak to her is afar-off guilty" (1.2.106)—and particularly from any further pernicious influence over her son, an association not lost on her: "My second joy,/ And first fruits from my body, from his presence/ I am barred, like one infectious" (3.2.94–96).[49] It is Leontes, however, whose effect on Mamillius has the most plaguey resonance, one that is revealing of this play's construction of the mind as a fragile thing, susceptible to infection from the exterior world.

Mamillius's death, so rapid and unexpected, an offstage happening, has been difficult to fathom for many critics. Those, such as Susan Snyder, who have attempted to diagnose the cause of his sudden demise usually associate it with his mother: symbiotically conjoined to the maternal body and breastfeeding through his name (an invention of Shakespeare's) "implied in its root *mamilla* (meaning 'breast' or 'teat')," Mamillius suffers a fatal lack of feminine nurturing when Hermione is removed from his presence.[50] While there are strong connections in the play between maternal influence and Mamillius's mortality, the impact of Leontes's actions on his son should not be underestimated. He is, after all, present during that fateful moment when Leontes observes Polixenes and Hermione together, presumably hears his father's initial feverish accusation and is subject to Leontes's all-too-hasty leap to the conclusion that perhaps even his first-born might not be his: "Mamillius,/ Art thou my boy?" (1.2.121–122). Then, directing his son, "this boy" that is so "like me" (1.2.137), to "Look on me with your welkin eye" (1.2.138), he dilates the progress of his cuckoldry, explains "the infection of [his] brains" (1.2.147), and implants an indelible image of his mother, "thy dam" (1.2.139), as a sexually false tormentor of the mind: "Thy mother plays" (1.2.188). In response to Leontes's rage against all "revolted wives" (1.2.200) with "no barricado for a belly" (1.2.205) as symptomatic of an epidemic "disease" that "Many thousand on's have" (1.2.207–208), Mamillius can only meekly reply: "I am like you they say" (1.2.209). After a brief appearance in which the boy, the audience learns, "troubles" (2.1.1) his mother, seems preoccupied with cosmetics and what he can learn "out of women's faces" (2.1.13), and tells "a sad tale's best for winter" (2.1.27), he is not seen again. The cause of his sickness and eventual passing are described unequivocally in terms of a conception in his mind:

> Conceiving the dishonour of his mother
> He straight declined, drooped, took it deeply,
> Fastened and fixed the shame on't in himself;
> Threw off his spirit, his appetite, his sleep,
> And downright languished. (2.3.13–17)

His death is reported by a servant who declares that: "The prince your son, with mere conceit and fear/ Of the Queen's speed, is gone" (3.2.142–143).

Later, Paulina indirectly blames Leontes for Mamillius's death, for planting injurious thoughts, jealous "fancies," in the child's mind:

> Whose honourable thoughts—
> Thoughts high for one so tender—cleft the heart
> That could conceive a gross and foolish sire
> Blemished his gracious dam. (3.2.193–196)

Mamillius thus dies of a conceit, an impression in his mind that produces an infected imagination and the fatal illness in his body.

As Snyder has shown, both *conceit*, "a (morbid) affection or seizure of the body or mind," and *conceive*, "used for the generation of a physical illness," were terms connected to notions of disease and embodiment at this time, but the idea that the force of the imagination could have transformative, even deadly, effects also had currency in contemporary thinking.[51] In such understandings, the mind could infect other minds. In "On the power of the imagination," Michel de Montaigne catalogues the kinds of miraculous happenings bred by the force of the mind alone but summarizes that:

> all this can be attributed to the close stitching of mind to body, each communicating its fortunes to the other. It is quite a different matter that the imagination should sometimes act not merely upon its own body but on someone else's. One body can inflict illness on a neighbouring one (as can be seen in the case of the plague, the pox and conjunctivitis which are passed on from person to person). . . . Similarly when the imagination is vehemently shaken it sends forth darts which may strike an outside object.[52]

In Montaigne's formulation, Leontes is the basilisk infecting Mamillius's "welkin eye" and impressionable mind with darts from his "shaken" imagination. Bacon also addresses the force of the imagination as contagion in *Sylva Sylvarum* but contends that the infection can be resisted:

> For as in *Infection*, and *Contagion* from *Body* to *Body*, (as the *Plague* and the like,) it is most certaine, that the *Infection* is receiued (many times) by the *Body Passiue*, but yet it is the *Strength*, and good *Disposition* thereof, Repulsed, and wrought out, before it be formed into a *Disease*; so much more in impressions from *Minde* to *Minde*, or from *Spirit* to *Spirit*, the impression taketh, but is Encountred, and Ouercome, by the *Minde* and *Spirit*, which is *Passiue*, before it worketh any manifest *Effect*.[53]

Contagious diseases do not infect everyone, just as not everyone is infected by disturbed imaginations like Leontes's, if their bodies and minds are strong enough to resist but Bacon concedes that contagions of mind are more likely to "worke most upon *Weake Mindes*, and *Spirits* [such] as those of . . . *Children* and *Young Creatures*."[54] Mamillius's tender age makes him particularly susceptible to taking on impressions. The preposterous conceit put into his mind by his raving father, presumably while looking directly into his eyes, is taken "deeply" into his imagination, producing an emotional infection powerful enough to kill.

In the opening scenes of *The Winter's Tale*, Shakespeare presents his audience with an extreme but plausible version of the kind of psychosomatic illness the early moderns were warned about in treatises about health, disease, and the emotions. The cry to constancy, to "a heart and mind so well settled and perfectly stayed," steadfast and resolute against inordinate motions "without any kinde of alteration or mutability," as Lemnius warned in *The Touchstone of Complexions* (1576), was the only bulwark against being "throwne into sundry diseases and innumerable affections . . . (like a Ship fraught with wares in tempestuous and boisterous weather) carried and dashed upon the rocks of perturbation."[55] Whatever healing may happen in the play's final acts through the restorative image, literally, of Hermione's statue coming to life, one that may even propose, against prevailing antitheatrical sentiment, that the theatre can offer as many curative as harmful images, Leontes's destructive passion and, particularly, the death of Mamillius reveal the tragic fragility of the early modern mind to contagions of the imagination.

NOTES

I am very thankful for the encouragement and judicious feedback received from Laurie, Lyn, and John on the draft version of this chapter, to my fellow participants in the "Body-Mind" Seminar at the 9th World Shakespeare Congress, Prague, 2011, for a robust discussion and to the Faculty of Arts at the University of Southern Queensland for funding my attendance at this event.

1. John Dover Wilson, *New Cambridge Shakespeare: The Winter's Tale*, ed. A. T. Quiller-Couch and J. D. Wilson (Cambridge: Cambridge University Press, 1931), 131.
2. Roger J. Trienens, "The Inception of Leontes's Jealousy in *The Winter's Tale*," *Shakespeare Quarterly* 4 (1953): 321–326; esp. 326.
3. J. L. Styan cited in Laurence Wright, "When Does the Tragi-Comic Disruption Start?: *The Winter's Tale* and Leontes's 'Affection'," *English Studies* 3 (1989): 225–232; esp. 225n4. Wright suggests that it is more important to focus on the moment when "Leontes's rational nature collapses" in the "affection" speech, rather than the question of when his jealousy is first conceived (227). His paper shares a conviction that Timothy Bright's *A Treatise of Melancholy* is helpful in providing a contextual understanding of Leontes's medical condition.
4. See Michael D. Bristol, "In Search of the Bear: Spatiotemporal Form and the Heterogeneity of Economies in *The Winter's Tale*," *Shakespeare Quarterly* (1991): 145–167; esp. 154; David Hillman, *Shakespeare's Entrails: Belief, Scepticism, and the Interior of the Body* (Basingstoke: Palgrave Macmillan,

2007), 155; Valerie Traub, *Desire and Anxiety: Circulations of Sexuality in Shakespearean Drama* (London and New York: Routledge, 1992), 44; and Gail Kern Paster, *The Body Embarrassed: Drama and the Disciplines of Shame in Early Modern England* (Ithaca: Cornell University Press, 1993): 260–270.

5. All citations of William Shakespeare's *The Winter's Tale* refer to *The Norton Shakespeare: Based on the Oxford Edition*, ed. Stephen Greenblatt, Walter Cohen, Jean E. Howard, and Katharine Eisaman Maus (New York: Norton, 1997).
6. See, for example, Michael O'Connell, *The Idolatrous Eye: Iconoclasm and Theatre in Early Modern England* (New York and Oxford: Oxford University Press, 2000), 139–142; and Marion F. O'Connor, "'Imagine Me, Gentle Spectators': Iconomachy and *The Winter's Tale*," in *A Companion to Shakespeare's Works,* Volume IV: *The Poems, Problem Comedies, Late Plays*, ed. Richard Dutton and Jean E. Howard (London: Blackwell, 2007): 365–388. See also James A. Knapp, "Visual and Ethical Truth in *The Winter's Tale*," *Shakespeare Quarterly* 55 (2004): 253–278. Knapp offers a rare consideration of the visual in the opening scenes of this play, drawing an enlightening connection between two key moments of seeing: Leontes watching the "entertainment" provided by Hermione and Polixenes and the revelation of Hermione's statue in the final scene (262).
7. Donald Beecher, "Windows on Contagion," in *Imagining Contagion in Early Modern Europe*, ed. Claire L. Carlin (Basingstoke: Palgrave Macmillan, 2005): 32–46; esp. 32.
8. Cited in E. K. Chambers, *The Elizabethan Stage: Volume 2* (Oxford: Clarendon, 1923), 340.
9. As Knapp states: "*The Winter's Tale* established in the early scenes the error of Leontes's interpretation. Nevertheless, because the text does not reveal what precisely it is that Leontes sees—but only that he sees differently from other characters—we are forced to contemplate the consequences of Leontes's possible reactions to what is apparently visible. The effect is to shift the attention from the spectacle onstage to the spectacle of Leontes's mind" (253).
10. See Stephen Orgel, "The Poetics of Incomprehensibility," *Shakespeare Quarterly* 42 (1991): 431–437; esp. 433–434.
11. Hallett Smith, "Leontes's Affectio," *Shakespeare Quarterly* 2 (1963): 163–166.
12. David Ward, "Affection, Intention, and Dreams in *The Winter's Tale*," *Modern Language Review* 82 (1987): 545–554; esp. 545–547.
13. Ward, 549.
14. Andre Du Laurens, *A Discourse of the Preseruation of the Sight: Of Melancholicke Diseases; of Rheumes, and of Old Age*, trans. Richard Surphlet (London, 1599), E1r.
15. Stuart Clark, *Vanities of the Eye: Vision in Early Modern European Culture* (Oxford: Oxford University Press, 2007), 1–2.
16. Clark, 10–11.
17. See Clark, 42.
18. As Clark suggests, "it takes only a moment's reflection to realize that a cognitive model which assumed that visual appearance would normally correspond with objective reality was ill-equipped to deal with serious and repeated breakdowns of that particular relationship—and more especially with situations where appearances that were supposed to be true proved difficult, perhaps impossible, to distinguish from appearances that were deemed to be false: let us call these 'visual paradoxes'" (20).

19. Clark, 53.
20. Thomas Wright, *The Passions of the Minde in General* (London, 1604), L3v–L4v.
21. Timothy Bright, *A Treatise of Melancholy* (London, 1586), H2r.
22. Bright, Ir–v.
23. Bright, Iv.
24. Bright, H4r–v.
25. Bright, J2v–J3r.
26. Bright, J3v.
27. Bright, K2r–K3r.
28. *OED*, *fancy n.*1, 5.
29. For more on the early modern belief in the power of the eye, see Sergei Lobanov-Rostovsky, "Taming the Basilisk," in *The Body in Parts: Fantasies of Corporeality in Early Modern Europe*, ed. David Hillman and Carla Mazzio (London: Routledge, 1997): 195–217.
30. In his seminal study *The Evil Eye: An Account of this Ancient and Widespread Superstition* (New York: Bell, [1895] 1989), Frederick Elworthy notes: "It was commonly believed in England at the time of the Black Death, that even a glance from the sick man's distorted eyes was sufficient to give the infection to those on whom it fell" (34).
31. Levinus Lemnius, *The Secret Miracles of Nature* (London: Early English Books Online, 1658), A2r–v.
32. See Francis Bacon, *Sylva Sylvarum: or, A Naturall Historie in Ten Centuries* (London: Early English Books Online, 1627), 247–248; and George Hakewill, *The Vanitie of the Eie* (London: Early English Books Online, 1608), 33–35.
33. Hakewill, 10.
34. For a detailed analysis of the significance of Hakewill's work see Clark (25–31), who defines it as "the most thoroughgoing denigration of vision in sixteenth- and seventeenth-century English, if not European thought" (25).
35. Beecher, 33.
36. Beecher, 37–39.
37. For an extended reading of this passage and its broader implications for the connections between antitheatricality, eye contagion, and lovesickness in *Twelfth Night*, see Darryl Chalk, "'To Creep in At Mine Eyes': Theatre and Secret Contagion in *Twelfth Night*," in *Rapt in Secret Studies: Emerging Shakespeares*, ed. Darryl Chalk and Laurie Johnson (Newcastle upon Tyne: Cambridge Scholars, 2010): 171–194.
38. Jacques Ferrand, *A Treatise on Lovesickness*, trans. and ed. Donald A. Beecher and Massimo Ciavolella (New York: Syracuse University Press, 1990), 252.
39. Linda Woodbridge, *The Scythe of Saturn: Shakespeare and Magical Thinking* (Urbana and Chicago: University of Illinois Press, 1994), 222.
40. Woodbridge, 222.
41. Robert Burton, *The Anatomy of Melancholy*, ed. Holbrook Jackson (London: J. M. Dent & Sons, 1932), 257.
42. Benedetto Varchi, *The Blazon of Jealousie* (London: Early English Books Online, 1615), C3r, H1r and E1v.
43. Varchi, C2r.
44. *Tell-Trothes New-yeares Gift* (London, 1593), D4v.
45. All citations of Ben Jonson's *Every Man in his Humour* are to the New Mermaids 2nd edition, ed. Robert N. Watson (London: A&C Black, 1998).
46. In his introduction, Watson speculates that Shakespeare, "listed first in the cast," may have played Knowell "first in the *Dramatis Personae*, though it is

tempting to suppose that he instead played Kitely, who was called Thorello in the early version of the play and is strikingly echoed by Shakespeare's Othello a few years later"—we must surely see similar echoes in Leontes.
47. Burton, 263–264.
48. John Downame, *Spiritual physicke to cure the diseases of the soule, arising from superfluitie of choller* (London: Early English Books Online, 1600), 58.
49. On the ways in which quarantine figured in Shakespearean drama, see Paula S. Berggren, "Shakespeare's Dual Lexicons of Plague: Infections in Speech and Space," in *Representing the Plague in Early Modern England*, ed. Rebecca Totaro and Ernest B. Gilman (New York and Oxford: Routledge, 2011): 150–169. Berggren argues that the frequent use of enclosed spaces and enforced isolation in Shakespearean drama would have brought to mind the regular measures used in plague time for the early modern spectator: "In Shakespeare's verbal lexicon of plague, the word *quarantine* is never spoken, but works like *Othello*, *The Winter's Tale*, and, most explicitly, *Romeo and Juliet*, dramatize the profound impression that isolating the sick seems to have made on the English populace" (150).
50. Susan Snyder, "Mamillius and Gender Polarisation in *The Winter's Tale*," *Shakespeare Quarterly* 50 (1999): 1–8; esp. 4.
51. Snyder, 4.
52. Michel de Montaigne, *The Complete Essays*, trans. M. A. Screech (London: Penguin, 1993), 118.
53. Bacon, 242–243.
54. Bacon, 243.
55. Levinus Lemnius, *The Touchstone of Complexions* (London: Early English Books Online, 1576), H4v and B2v.

7 Singularity in *The Winter's Tale*

Hardin Aasand

> A poetics of allegory can achieve stability only by grounding itself on an ambiguous determination of the "other" in the word *allegory* as that to which the discourse refers, the ideal "meaning." This determination is formalized in poetics as what I call the *singularity*, the ineffable presence into which, it is supposed, everything in the allegorical work ultimately is drawn.
>
> —Gordon Teskey, *Allegory and Violence*[1]

> Your gallery
> Have we passed through, not without much content
> In many *singularities*; but we saw not
> That which my daughter came to look upon,
> The statue of her mother.
>
> —*The Winter's Tale* (5.3.10–14)[2]

When he walks into Paulina's chapel, accompanied by his long lost daughter Perdita, Leontes utters a word rife with meaning in the seventeenth century: *singularity*, a word layered with suggestions of isolation, uniqueness, aesthetic curiosity. It is a suggestive word for a man whose perverse rejection of spousal fidelity, filial familiarity, and prophetic utterance had isolated him in his own claustrophobic remove during the first half of the play, a "psychological projection" in Howard Felperin's terms, that contaminated all ostensible objectivity with a delusional certainty:[3]

> How blest am I
> In my just censure! in my true opinion!
> Alack, for lesser knowledge! How accurs'd
> In being so blest! (2.1.36–39)

The transparency of Leontes's rhetorical certitude here is met by the opacity of subsequent justifications defined by syntactical solecisms and incomprehensible referents, a cognitive black hole. Leontes infers so much when he counters Hermione's claims of innocence by characterizing his evidentiary core as incapable of balancing a "school boy's top" should his fears be found false. His core threatens to collapse upon itself. In Stephen Orgel's words, Leontes falls into a "poetics of incomprehensibility."[4] In the course of the play's five acts, Leontes exchanges this fragile delusional state for one that Shakespeare invests with wholeness and auratic profundity. In terms of cognitive theory, Leontes's transformation reflects the dissolution of his hermetic boundaries and his return to an embedded—and subsequently extended—cognitive presence in Sicilian society. As Evelyn Tribble and Nicholas Greene assert, the notion of "extended mind" or "distributed cognition" integrates the individual cognitive practice within the symbols and artefacts that culture provides us for processing human understanding:

> Both Extended Mind theory and Distributed Cognition posit that the mind is both embedded in and extended into its worlds. Rather than seeing 'cognition' as a set of interior trans-historical mental processes, extended and distributed models hold a hybrid and integrative view, in which the realm of the mental extends into physical and social systems.[5]

Thus Leontes's exculpatory reemergence as a vital, whole parent and husband demands this reincorporation of his cognitive engagement within an ecosystem redeemed by and distributed through forgiveness and familial restoration.

Walking through Paulina's gallery, Leontes confronts a statue of the mother he banished and whom Perdita has never seen—a spouse he had cruelly dispatched to prison and memorialized for sixteen years through a ritual of hermetic "recreation." For Leontes, this singularity is a statue designed for veneration, a rarity of Paulina's creation that, like a lodestone, draws him and his daughter to the play's conclusion. The isolation of Paulina's gallery—its investment with a sacred aura—provides the unveiling of the statue with a numinousness befitting singularities, a hollowing out of a space to situate them as distinctive, solitary, and sacrosanct. As defined by Gordon Teskey in his study of allegory, the singularity is analogous to the "vanishing point" in a linear perspective: invisible yet foundational in creating proper vision and depth of field for visible phenomena. Singularities are essential for the functioning and sustenance of allegory; equally important, however, is the function of singularity for engendering a heightened wisdom marked by the violent "capture" of meaning from the "noise" of life. Such wisdom is found within Paulina's chapel.

In *Christian Morals*, a meditative exposition written in the 1670s but published posthumously in 1716, Thomas Browne concludes that the proper comprehension of Time based on the cumulative experience of "singularities" is richly productive of wisdom:

> In seventy or eighty years a Man may have a deep Gust of the World, Know what it is, what it can afford, and what 'tis to have been a Man. Such a latitude of years may hold a considerable corner in the general Map of Time; and a Man may have a curt Epitome of the whole course

> thereof in the days of his own Life, may clearly see he hath but acted over his Fore-fathers, what it was to live in Ages past, and what living will be in all ages to come.
>
> He is like to be the best Judge of Time who hath lived to see about the sixtieth part thereof. Persons of short times may Know what 'tis to live, but not the life of Man, who, having little behind them, are but Januses of one Face, and Know not singularities enough to raise Axioms of this World: but such a compass of Years will show new Examples of old Things, Parallelisms of occurrences through the whole course of Time, & nothing be monstrous unto him; who may in that time understand not only the varieties of Men, but the variation of himself, and how many Men he hath been in that extent of time.[6]

As personified by the god Janus—Time—truth and wisdom are effected through a paradoxical process that is at once formative and summative, a standing within and a standing outside of mortal existence, the apprehension of life's *saeculum* in a totality arising asychronously from life itself. Janus's two faces convey Time as both *chronos* and *kairos*; that is, both as operational and linear and as numinous and proleptic. As Achsah Guibbory observes in her study of Thomas Browne, "[The] skeptical Janus represents Browne's conception that the truest wisdom a man can possess on earth is an inclusive view of time which recognizes that full knowledge can only exist beyond this temporal world"[7] For Browne, Janus is a "mortifying presence" to be held in one's mind and not relegated to ornamental "signets" displaced to the hand. Wisdom advances concomitantly and paradoxically both within time and outside of time. As emblematized by Don Sebastian de Covarrubias—Browne's own marginal memento mori—Janus's two faces are a two-sided reversal of Cartari's conventional Janus emblem (the "withered & hoary" past and the "youthful & beardless" future): "two Faces averse and conjoined Janus-like, the one a Gallant Beautiful Face, the other a Death's Head Face" replete with the Ovidian motto, *Quid fuerim quid simque vide.*[8] For Browne, this reversal in Janus's visage portends a future marked not by life *in potentia* but rather by a promissory note of eschatological opportunity heralded by memento mori: linear *chronos* gives way to a memento mori that hints discretely at a kairotic ascendancy, "the saturated time of opportunity, where the *not yet* urgently arrives."[9]

In this essay, I hope to move the conversation of Time's appearance in *The Winter's Tale* beyond the merely functional role this choric throwback performs to compensate for a sixteen-year gap in the play to Shakespeare's aesthetic use of personification to interrupt the play's linear time structure. More than a device for compressing the narrative into a dramatic form accessible for a theatre audience, Time erupts into the play to construct a dramatic "vanishing point," an untimely singularity annunciated by the overt anachronism of allegory. As Alexander Nagel and Christopher Wood observe, the deliberate intervention of anachronism, of *spolia* or fabricated antiquities in Renaissance arts, creates an "omnitemporal" space that "lift[s] the event out of the flow of history," embedding within the aesthetic medium a "mystery, figure, or sacrament" that defines a temporal, troped singularity.[10] This singularity, as Teskey reminds us in his study of allegory, is an allegorical manifestation of a feigned coherence that humans establish against the onslaught of meaningless chaos, the cosmic noise of an unordered agon emanating from allegory's genealogical belatedness: "At the most fundamental level genealogy is expressed in a savage, Hesiodic narrative in which the happiness of the Olympians emerges from a history of cannibalism, mutilation, and capture."[11] We can detect a Shakespearean appropriation of this paradigm in Leontes's penitential seclusion to expiate the irrational jealousy that wreaked the carnage of the play's first half: dismantled familial order and disrupted genealogical succession. Out of this "mutilation" emerges Time's allegorical presence.

For A. D. Nuttall, however, Time rates no mention: the play is "broken-backed" and the transition from tragic demise to pastoral hopefulness is qualified as a "violence of division," the musically overlapping tragic end of Antigonus sounded rustically in the emergent pastoral tones of the clown's description of the shipwreck and carnivorous removal of Antigonus: "A lurching, earthy rustic dance, with a bear and a gentleman that seem to have stepped out of a nursery-rhyme book, is afoot before the ascent into happiness has properly begun."[12] For Nuttall, who delegates Time's role to the clown, Act 4, Scene 1 is an occluded element. The clown effects the transition through his elegiac narration of Antigonus's demise. Michael Bristol conceives of Time as a functional pivot in Shakespeare's manipulation of temporal experience. Though Bristol brings it into his analysis, the choric Time is relegated to a cursory, functional role that foregrounds the "derangement" in the "spatiotemporal heterogentity" of the play's "chronotype": "The chorus openly admits here that the passage of time is not only without duration in the ordinary meaning of that concept but also without content. Sixteen years are missing, and for all intents and purposes they are empty."[13] Thus, for Bristol, Time is a chorus who represses the "growth" in an untried, unrepresented format, a rupture that Time explicitly acknowledges as the play moves dramatically from the wintry Candlemas bear to a Lenten resurrection of hope within a Bohemia defined by commodity and market forces.

William Blissett and Inga-Stina Ewbank discuss more explicitly the allegorical presentation of Time: for Blissett, Time is "creaky and fusty," atavistic in effect[14]; for Ewbank, Time clearly derives from Robert Greene's *Pandosto*, the prose romance that clarions the "Triumph of Time" through an elaborate argument and epigraph: "although by the means of sinister fortune Truth may be concealed, yet by Time in spight of fortune it is most manifestly revealed. Time, Temporis filia veritas, whose daughter is Truth."[15] As Ewbank observes, however, Greene's prose foregrounds Fortuna, who, though subject to Time, manipulates the story of Pandosto to engender suicide and truncated happiness. Ewbank usefully reminds us that Shakespeare drew on the iconographic figure of Father Time adumbrated in a number of pageants and masques in 1610–1611, in which Time as Devourer/Revealer is given iconographic presence. While she provides an iconographic and symbolic precedent for the appearance of Time in Act 4, Scene 1, Ewbank does not scrutinize the assumptive inclusion of allegory as a choric figure and the etiology behind Shakespeare's decision to revert to a medieval allegory for the play's narrative bridge. Why transform the chorus into a prosopopoeia of Time?

The allegorical figure of Time itself takes center stage both to expedite Leontes's rehabilitation and engender a pastoral mode that transmutes the tragic genre of the first three acts. Furthermore, allegory itself is not immune from its own allegorical potentia: in opening a chronological rift in the emerging tragedy prompted by Leontes's tragic inclination, Time transmutes both the site of the play and its generic modality, exchanging Bohemia for Sicilia and tragedy for tragicomedy. Allegorical structures reclaim order from the threat of narrative dissonance, as Teskey observes: "The very word *allegory* evokes a schism in consciousness—between a life and a mystery, between the real and the ideal, between a literal tale and its moral—which is repaired, or at least concealed, by imagining a hierarchy on which we ascend toward truth."[16] In *The Winter's Tale*, Time "repairs" Leontes's contaminating delusions by reestablishing a generic and temporal distance through Ovidian allusions and pastoral sublimation that refigure and recycle a past haunted by death and familial dissonance. Dramatically, Time performs a cognitive function of extending the audience's understanding by extending the symbolic system of the stage through an intentional anachronism. Time is Shakespeare's means of distributing meaning and cognitive processing for a restorative closure.

Time's spatiotemporal position in the play is itself significant: it follows the discovery of the abandoned infant, Perdita, and the clown's rich description of both a tempest that "flap-dragoned" Antigonus's departing ship and a coincidental dismemberment of Antigonus by the wayward bear who swallows the courtier as the ship is swallowed by the sea.[17] Jonathan Bate notes that Ovidian allusions emanate here and proliferate through the remainder of the play.[18] Thus it is an Ovidan bear—Callisto, a victim of jealousy metamorphosed into this ursine form—who devours the pathetic Antigonus. Even more compelling, Time appears only after the older shepherd and clown decide to bury Antigonus's remains. Time emerges from the corporeal remains of Antigonus's flesh to swallow up the untried growth in one scene, an appropriate origin for allegory: "Allegory has its origin in that process of hierarchical ordering which is in reaction to the chaos of Saturnian, allelophagic ["a mutual devouring" of forms] desire."[19] For Teskey, allelophagy is the mutual devouring of bodies engaged in a battle to enclose the other, the basis for the instrumental meaning that provides for allegorical foundation. All allegory is thus an act of latent violence that undergirds all attempts to construct meaning in an allegorical sense. Time is incarnated following the bear's feast.

It is equally fitting for Shakespeare to harken back to the medieval stage to resurrect Time, for the deliberate archaic nature of Time's presentational style—Time does not "represent" but "presents"—reverts to a more labile theatrical experience, as Robert Weimann has demonstrated in *Author's Pen and Actor's Voice*, in which the presenter occupied both the engaged *platea* and the symbolic *locus* of the stage.[20] As K. Janet Ritch notes, the "liminal presenter" was the "double-faced Janus" of medieval stage, a figure "poised upon the threshold between fact and fiction, reality and illusion, history and contemporary society."[21] The fixing of an allegorical figure within the play's diptych structure establishes a transition that not only bridges but acknowledges the "vast" covered between the sections of the play. Indeed, the Janus nature of Time allows for a simultaneity of "momentary skepticism and contemplation." The audience recognizes the personification, but Shakespeare's audacity in giving symbolic form to Time's abstract existence is dramatically alluring. Just as Paulina reorients Leontes's own cognitive process of remembering and seeking penance through an enforced "recreation" for his repressive treatment of Hermione, Shakespeare recalibrates his audience's own experience by the allegorical interruption of generic expectation. Ellen Spolsky has characterized such jarring disruptions of cognitive expectations as "grotesque,"[22] and there is a sense in which Time's appearance indeed acquires a patina of the grotesque as a liminal presence that serves as a semiotic radical above the narrative ostensibly containing his presence.[23]

When he arrives on the shores of Bohemia, Antigonus enters into a world that is geographically and anachronistically distant from the tragedy of Sicilia but by no means immune from tragic potential. It is a world in which nature impinges ethically on man's actions:

> Ay, my lord, and fear
> We have landed in ill time: the skies look grimly
> And threaten present blusters. In my conscience,
> The heavens with that we have in hand are angry
> And frown upon's. (3.3.2–6)

The Mariner's attempt to associate present action with a divine displeasure is not the first time such an interpretive gesture occurs. Leontes, likewise, in isolated fashion reads the concomitant deaths of Mamilius and Hermione through the belated lens of his own baseless jealousy. Such retribution gives shape to Leontes's groundless suspicions: "Apollo's angry, and the heavens themselves/ Do strike at my injustice" (3.2.143–144). His distress takes the form of a direct invocation of Apollo's forgiveness for his gainsaying the Delphic oracle: "Apollo, pardon/ My great profaneness 'gainst thine oracle" (3.2.150–151). Both of these responses treat natural forces as sources of retribution for unpardonable actions each has committed, but the mariner's conclusion invests all of Bohemia with immanent moments of divine intervention that are at once atavistic and resonant with theophanic possibilities befitting the pastoral mode.

Antigonus's guilt is further embodied in the recounting of his dream of Hermione. The apparition who visits him is a ghostly precursor of the statue that will release Leontes from his penitential isolation. Antigonus's dream is the first such uncanny resurrection of Hermione, an intrusive, apotropaic presence that transfixes him. Significantly, however, Antigonus misreads this nocturnal visit and finds in it only a further reification of Leontes's adulterous slander:

> I have heard, but not believed, the spirits o'th' dead
> May walk again. If such thing be, thy mother
> Appeared to me last night, for ne'er was dream
> So like waking. To me comes a creature,
> Sometimes her head on one side, some another;
> I never saw a vessel of like sorrow
> So filled and so becoming. In pure white robes
> Like very sanctity, she did approach
> My cabin where I lay; thrice bowed before me,
> And, gasping to begin some speech, her eyes
> Became two spouts. (3.3.15–25)

Hermione's words are part of the "tragic economy" that is subterranean and omnipresent, even in the pastoral Bohemia. This spectral presence and its emotive gesture enthrall Antigonus, riveting him and eliciting from him a "squaring" of the historical record: the dream must accord with the slanderous abuse that had become Leontes's and by extension Antigonus's reality. In a sense, Hermione's apparition, even filtered through and misread as spectral confirmation of Leontes's fears, is simultaneous with Antigonus's death:

> Affrighted much,
> I did in time collect myself, and thought
> This was so, and no slumber. Dreams are toys,
> Yet for this once, yea superstitiously,
> I will be squared by this. I do believe
> Hermione hath suffered death, and that
> Apollo would, this being indeed the issue
> Of King Polixenes, it should here be laid,
> Either for life or death, upon the earth
> Of its right father. (3.3.36–45)

The appearance of both a misread and misprized Delphic oracle and a ghostly apparition prepares the play for Time's unexpected and uncanny appearance. Time and its allegorical remove provide for the "temporal and spatial mediation"[24] that reveals the status of all oracular truths to be "tropic movement[s]" present in narrativity, the expanse of time that engenders chronological and spatial distance. The inability of Leontes and Antigonus to trust respectively the oracular or the uncanny dreamwork of apparitions is predicated on an emotional proximity that lacks the distance and objectivity by which a thoughtful *poesis* or genre makes use of the past. As Teskey notes, "If . . . we look behind the term *genre* to its root in genesis and genealogy, we discover a more violent scene: the perennial assault of the generated on what is established before it in time, on that which has given it substance."[25] The Ovidian allusion of Callisto that subtends Antigonus's bear reflects Shakespeare's own belated interweaving of order on the play's random violence, a figural palimpsest that refigures and reshapes the chaos of tragic collapse.

When it appears on stage, Time authorizes Shakespeare's appropriated surrogacy. The play's unresolved skeins of narrative (Hermione's apparent demise, Leontes's promise of devotional visitation to the collective grave of mother and son, the banished Perdita separated by sea from her family) are fertile ground for allegorical intrusion, a Spenserian revenant that both cleaves and fuses the play's "broken-back" structure. The clown's description of the tempest that marks Antigonus's remove evokes the chaos that has engulfed all natural order and collapsed boundaries: "I have seen two such sights, by sea and by land! But I am not to say it is a sea, for it is now the sky; betwixt the firmament and it you cannot thrust a bodkin's point." Bate describes this account as Ovidian in force, as we have seen, and one can indeed feel the pulse of it from Book 1 of Ovid's *Metamorphoses*:

> For where was earth, was sea and ayre: so was the earth unstable,
> The ayre all darke, the sea likewise to beare a ship unable,
> No kinde of thing had proper shape, but ech confounded other.
> For in one self same bodie strove the hote and cold togither,
> The moyst with drie, the soft with hard, the light with things of weight.
> (1. 15–19)[26]

The natural pandemonium echoes the impassioned madness that defines Leontes's own *tremor cordis*, an emotional collapse in which trusted "friendship" and the intimate "blood" of marriage mingle in frenzied fashion, rendering Leontes distraught and unsound. But just as he presents us with a nature out of control, one that devours a ship and cannibalizes Antigonus, Shakespeare reclaims an equilibrium through the spontaneous, eulogizing gesture of commemoration. Antigonus's dismembered, shapeless remains are reconstituted and interred within a grave that imparts form and reconstitutes matter.

Time is also prepared for elliptically in the clown's disparagement of the ages between ten and twenty-three. Just as Time passes over the "untried growth" of sixteen years, the clown regrets the activities that define youth: "there is nothing in the between but getting wenches with child, wronging the ancientry, stealing, fighting" (3.3.58–62). If only "youth would sleep out the rest." With Time's arrival, the play enacts just such a winking out of growth. Time's monologue to the audience is as riven as the play's structure. Initially, there is an iconographic, authoritative force to the speech as Time establishes its prerogative and its emblematic sway over nature:

I that please some, try all; both joy and terror
Of good and bad, that makes and unfolds error,
Now take upon me, in the name of Time,
To use my wings. . . .
Let me pass
The same I am ere ancient'st order as,
Of what is now received. I witness to
The times that brought them in; so shall I do
To the'freshest things now reigning, and make stale
The glistering of this present as my tale
Now seems to it. (4.1.1–15)

This primordial state positions Time as both inside and outside of order, a fluency with both states that compels allegory to portray its capacious nature. With the turn of the emblematic hourglass, the address is transmuted into a distinct, narrative account of diurnal existence. Time figuratively domesticates itself:

Your patience this allowing,
I turn my glass, and give my scene such growing
As you had slept between; Leontes leaving—
The 'effects of his fond jealousies so grieving
That he shuts up himself—imagine me,
Gentle spectators, that I now may be
In fair Bohemia. (4.1.15–21)

While Perdita's name originates from Hermione's awareness that she is "lost," her future husband, Florizel, is given identity by Time, a move that marks the play's divergence from Greene's *Pandosto*, where Florizel is named Dorastus. Time grafts Ovid into the play through this namesake, implanting a fertile, mythic allusion that both enriches and complicates Florizel's and Perdita's romance. Time's coyly elliptical avoidance of Perdita's story ("What of her ensues/ I list not prophesy") urges the story forward, interweaving Perdita with her Ovidian counterpart Proserpina (and by extension, Persephone).

If it is configured as a Janus bifurcated by past and future, Time is also bifurcated by gender: "The choric figure of Time . . . could readily be female as the birth-giver to event," as Helen Hackett notes, despite the self-reference to "himself" and the iconographical allusions that are patriarchal and male-gendered.[27] Even the iconographic record shows Time's ambiguous gender. An anonymous 1589 work, *Temporis Filia Veritas: A mery devise called The Troublsome travell of Tyme, and the daungerous delivery of her Daughter Trueth*, records Time's variable gender. In a postscript, the author neuters Time: "Behold euen through such a great Trouble and trauell aforesayd, will gentle Ladye Tyme, bring forth her most bewtifull Daughter Trueth."[28] Although Time's magisterial couplets seek to vanquish the past—to create the necessary lapse of sixteen years and overthrow custom—the conditions that Time voices are rife with fecundity and pastoral promise, both in terms of the narrative but also in terms of the Ovidian allusions that belatedly emerge through the imagery and the lyrical echoes. Thus, despite Time's assumption of a sweeping prerogative that rivals Leontes's own peremptory disposition in the first three acts—"Why, what need we/ Commune with you of this, but rather follow/ Our forceful instigation?" (2.1.162–164)—Time also engenders a growth that, despite the inclination to elide process, is productive of the "romance" form that the play inhabits.

If it promulgates generic shifts from tragedy to romance, Time also elicits the play's allusive pregnancy of a violent intertextual plenitude, a generative transformation borne from the fertile aftershocks of generational assaults on the past. The fecundity that Acts 4 and 5 celebrate through Florizel and Perdita—the promise of a genealogy emanating from them into the future—devolves from the violence that Time tamps down in Act 4, Scene 1: the death of Mamillius, the supposed death of Hermione, the drowning of all Antigonus's crew, and the devouring of Antigonus by a bear, which provides the first major Ovidian allusion in the play. In Bohemia, violence is ostensibly sublimated, restrained by the narrative from erupting as traumatically as it did in Sicilia. Even with Florizel, Ovid's presence constrains Florizel's incipient lust, visible in Apollo's reemergence in the allusive analogy Florizel constructs of divine metamorphosis:

> The gods themselves,
> Humbling their deities to love, have taken
> The shapes of beasts upon them. Jupiter
> Became a bull, and bellowed; the green Neptune
> A ram, and bleated; and the fire-robed god,
> Golden Apollo, a poor humble swain,
> As I seem now. (4.4.25–31)

Florizel's self-restraint—his sublimation of passion—is amplified and countered by the precipitous carnality woven into the subtle skeins of Arachne's tapestry, an Ovidian tale that recounts divine abuse of metamorphosis:

> The Lydian maiden in hir web did portray to the full
> How *Europe* was by royall *Jove* beguilde in shape of Bull.
> That Bull and Sea in very deede ye might them well have thought.
> . . . . .
> She also made *Neptunus* leaping by
> Upon a Maide of Aeolus race in likeness of a Bull . . .
> In likenesse of a Countrie cloyne was *Phebus* picturde there,
> And how he now ware Gossehauke's wings, and now a Lions heare.
> And how he in a shepeherdes shape was practising a wile
> The daughter of one *Macarie* dame *Issa* to beguile.
> (*Metamorphoses* 6.126–129, 141–155)

As Leonard Barkan concludes, Arachne's tapestry—woven by a woman to record mortal vulnerability to gods' sexual advances—demonstrates metamorphosis as a process of impassioned eros that finds its meaning only in aesthetic reproduction.[29] In Florizel, Shakespeare demonstrates both the urge and the human restraint of passion in the selfsame moment. If the result is awkward, it captures nevertheless the vagaries of life that threaten an order predicated on human deliberation and gentle courtesy.

While much has been written of Hermione's dramatic return to life—a "statue" carved from the textual weavings of multiple Ovidian tales, from Pygmalion to Orpheus and the Deucalion/Pyrrha stories—it is the reiterative power of Hermione to assert her singularity in this play, her capacity to become the seminal vanishing point of meaning for the play, that I want to comment on briefly. Time establishes a proleptic presence that compresses the maternal and the temporal; despite the promise of an "untried growth," Hermione's wrinkles incorporate the "great carver's" role in etching sixteen years into her face, and this aesthetic singularity is no less ineffable for the wrinkles. Its presence, like Keats's Grecian urn, records both an imperviousness to Time and a vulnerability to its temporal signature, a signature that communicates the emotional deprivation generated by her hermetic seclusion and abandonment. Transfixed by an ecstasy that ennobles Hermione's iconic wrinkles, Leontes and Perdita are left rapt by a marble statue susceptible as flesh to Time's touch:

> Does not the stone rebuke me
> For being more stone than it? O royal piece!
> There's magic in thy majesty, which as
> My evils conjured to remembrance, and
> From thy admiring daughter took the spirits,
> Standing like stone with thee. (5.3.37–42)

As Lowell Gallagher notes, it would be a simplistic—and errant—response to read this moment, as Polixenes and Camillo do, as a *tableau vivant* static and lifeless, or to treat it as a numinous theophany that arrests Leontes and Perdita and transforms them into immobile simulacra. Shakespeare ultimately entrusts Perdita and Hermione with the narrative unfolding that marginalizes Leontes. In her lines to Perdita, Hermione establishes the authorship of this singularity: "For thou shalt hear that I,/ Knowing by Paulina that the oracle/ Gave hope thou wast in being, having preserved/ Myself to see the issue" (5.3.125–128). This gives weight to the conclusion reached by Helen Hackett that *The Winter's Tale*, more than any of the other late romances, establishes the "fantastical, digressive and time-filling narrative principles of romance."[30]

Indeed, the "spilling over" of narrative that Leontes desires from Paulina reflects the surplus of narrative that time and generation nurture. Time, the choric figure of compressed linear chronology and "untried" growth, ironically extends the play beyond its margins. While Dennis Kay sees the collective removal from the stage as a promissory note for a full recounting to Leontes of the losses incurred by his jealousy—Antigonus, Mamillius, the severing of marital, filial, and fraternal bonds—one would be hard pressed to see Hermione's silence as merely a "tongue-tied" return to her earlier demure posture.[31] Despite the substitutions the play enacts—Camillo for Antigonus, Perdita for Mamillius—the promised order only cursorily refutes death through such compensatory substitutions. Perhaps Thomas Browne's own version of singularity is optimal: "a compass of Years" unfolds examples, parallels, and variations that are fashioned, like Arachne's tapestry or Hermione's wrinkled figure, into an eschatological singularity begetting wisdom. If the characters' concluding departure from the stage promises a narrative to be woven by Hermione's renewed silence and Paulina's lawful magic, the wisdom Leontes obtains arrives only after a stark remembrance that the wrinkles on Hermione's face record an absence etched by unkind jealousy and time's indulgence. As Lowell Gallagher notes, "More than merely visible, the wrinkles are legible as the signature of temporality. . . . The statue/ *imago* in Paulina's gallery, then, foregrounds the negotiations of containment and dispersal that characterize historical contingency and the very structure of the self in time."[32] And Time's own daughter is Truth (*Temporis filia veritas*), for whom no mother can be recovered and no statue erected.

EPILOGUE (NOT SPOKEN BY TIME)

The textual history of *The Winter's Tale* is fraught with absences and delays, as if Time the Chorus circumvented the textual history to "overthrow custom" in the printing house. Lori Newcomb provides a succinct account of the prevailing narrative regarding the missing "allowed book" and the apparent chaos in the production of the First Folio:

> To early textual scholars, it seemed more than coincidental that *The Winter's Tale* went missing from two different venues on or about

> August 1623. Could the use of the "allowed text" in the printing house have rendered it unavailable for performance licensing? Or conversely, could the 1623 performance of *The Winter's Tale* have temporarily derailed its publication? Alfred W. Pollard mused in 1907: "If only one manuscript of the play was available, and that was wanted for use in the theatre when the end of the comedies was reached, there may have been a delay in printing this one play . . . But we cannot be sure that we have anything more than a coincidence."[33]

The placement of the play in the Comedy section of the Folio is marked by the infamous blank verso page between the final recto page of *Twelfth Night* and the first recto page of *The Winter's Tale*. The reader of the Folio thus finishes a reading of Feste's concluding song, which extends the dynamics of that play out to the audience and reader. The song, which traces the span of human mortality, resonates with a refrain that announces, "For the rain it raineth every day." Feste's "whirligig of Time" governs not only this play but also that of the truant play that follows.

The intervening blank page marks an interruption, the disappearance of Time and a delay, both of which define the vagaries of Time's prerogative. More telling, perhaps, is the last page of the play, a bookend complementing the initial, blank verso page. This last page records the "Names of the Actors." Curiously, a close reading of the itemized list reveals that the only character missing from the roster is Time itself. Accident, oversight on the part of Ralph Crane (the scrivener for the text), or Time's own volition? Perhaps Time itself merely lacked time for such a formal device.

NOTES

1. Gordon Teskey, *Allegory and Violence* (Ithaca, New York: Cornell University Press, 1996), 5.
2. *The Winter's Tale*, ed. Susan Snyder and Deborah T. Curren-Aquino (Cambridge: Cambridge University Press, 2007). All citations are quoted from this edition.
3. Howard Felperin, "'Tongue-tied, O Queen?' The Deconstruction of Presence," in *The Winter's Tale. Shakespeare: The Last Plays*, ed. Kiernan Ryan (London: Longman, 1999): 187–205.
4. Stephen Orgel, "The Poetics of Incomprehensibility," *Shakespeare Quarterly* 42.4 (1991): 431–437.
5. Evelyn B. Tribble and Nicholas Keene, *Cognitive Ecologies and the History of Remembering* (New York: Palgrave Macmillan, 2011), 2.
6. *The Works of Sir Thomas Browne*, ed. Geoffrey Keynes. 4 vols. (London: Faber & Faber, 1964), I: 284.
7. Achsah Guibbory, "Sir Thomas Browne's Allusions to Janus," *English Language Notes* 12 (1975): 269–273; esp. 272.
8. Browne, I: 277.
9. Lowell Gallagher, "The Sacramental Neuter and the Missing Body in Robert Southwell's Poetics," in *Go Figure: Energies, Forms, and Institutions in the*

Early Modern World, ed. Judith H. Anderson and Joan Pong Linton (New York: Fordham, 2011): 38–57; esp. 43.

10. Alexander Nagel and Christopher S. Wood, "Interventions: Toward a New Model of Renaissance Anachronism," *Art Bulletin* 87.3 (2005): 403–415; esp. 408.
11. Teskey, 172. Teskey sees genealogy as the belated battle that threatens an "order of forms," a memory that threatens to overwhelm established authority. Also see Lowell Gallagher's "'This Seal'd-Up Oracle': Ambivalent Nostalgia in *The Winter's Tale*," *Exemplaria* 7.2 (1995): 465–498. Gallagher elaborates on belatedness as a "necessary condition of sign production" that "names the hollowed-out space between the literal and the figural, the space articulated and filled in by hermeneutic labor" (470).
12. A. D. Nuttall, *Shakespeare the Thinker* (New Haven, Connecticut: Yale University Press, 2007), 350.
13. Michael Bristol, "In Search of the Bear: Spatiotemporal Form and the Heterogeneity of Economics in *The Winter's Tale*," *Shakespeare Quarterly* 42 (1991): 145–167; esp. 146.
14. William Blissett, "'This Wide Gap of Time': *The Winter's Tale*," *English Literary Renaissance* 1.1 (1971): 52–70.
15. Inga-Stina Ewbank, "The Triumph of Time in *The Winter's Tale*," in *The Winter's Tale: Critical Essays*, ed. Maurice Hunt (London: Routledge, 1995): 139–155.
16. Teskey, 2.
17. See Patricia Parker's work on the "polymorphic bear" in the play. She notes that Antigonus's name means "antibearing or birth" and combines the homophonic play on "bears," "barne," and the Greek, "gone, generation." Antigonus is not only "anti-birth" in terms of his disposition of Perdita, but becomes himself borne away by a "bear." "Sound Government, Polymorphic Bears: *The Winter's Tale* and Other Metamorphoses of Eye and Ear," in *The Wordsworthian Enlightenment: Romantic Poetry and the Ecology of Reading*, ed. Helen Regueiro Elam and Frances Ferguson (Baltimore: Johns Hopkins University Press, 2005): 172–190.
18. Jonathan Bate, *Shakespeare and Ovid* (Oxford: Clarendon Press, 1993), 223.
19. Teskey, 14.
20. See Robert Weimann's elaboration on the use of the stage—the *locus* of ideological content and the *platea* of malleable interrogation of play ideology in his *Author's Pen and Actor's Voice: Playing and Writing in Shakespeare's Theatre* (Cambridge: Cambridge University Press, 2000).
21. K. Janet Ritch, "The Role of the Presenter in Medieval Drama," in *"Bring furth the pagants": Essays in Early English Drama Presented to Alexander F. Johnston*, ed. David N. Klausner and Karen Sawyer Marsalek (Toronto: University of Toronto Press, 2007): 230–270; esp. 230.
22. Ellen Spolsky, *Word vs Image: Cognitive Hunger in Shakespeare's England* (New York: Palgrave Macmillan, 2007), 128. Spolsky's notion of "grotesque" is especially provocative here. Citing Geoffrey Harpham's superlative work on the "grotesque," Spolsky provides a useful definition: "grotesques are disturbing because they 'call into question the adequacy of our ways of organizing the world, of dividing the continuum of experience into knowable particles' . . . Grotesques themselves, he recognizes, have no intrinsic meaning, but 'place an enormous strain on the marriage of form and content by foregrounding them both so that they appear not as a partnership, but as a warfare, a struggle.'" I submit that Time performs as a grotesque, a struggle between representative forms.

23. For "semiotic radical," see James Dougal Fleming's "Meanwhile, Medusa in *Paradise Lost*," *ELH* 69 (2002): 1009–1028. Fleming characterizes Medusa's presence in Book 2 of *Paradise Lost* as Milton's incorporation of a "semiotic radical," a discursive presence within the epic's narrative that serves as Milton's own authorial intrusion into the work; Medusa is an apotropaic surrogate for Milton.
24. Gallagher, "Oracle," 498.
25. Teskey, 162.
26. W.H.D. Rouse, ed. *Shakespeare's Ovid: Arthur Golding's translation of the* Metamorphoses (New York: W.W. Norton, 1961). All quotations from Ovid are cited from this edition.
27. Helen Hackett, "'Gracious be the issue': Maternity and Narrative in Shakespeare's Late Plays," in *Shakespeare's Late Plays: New Readings*, ed. Jennifer Richards and James Knowles (Edinburgh: Edinburgh University Press, 1999): 25–39; esp. 37.
28. Soji Iwasaki, "*Veritas Filia Temporis* and Shakespeare," *English Literary Renaissance* 3.2 (1973): 249–263; esp. 252.
29. Leonard Barkan, *The Gods Made Flesh: Metamorphosis and the Pursuit of Paganism* (New Haven, Connecticut: Yale University Press, 1990), 2–5.
30. Hackett, 36.
31. Dennis Kay, "'To hear the rest untold': Shakespeare's Postponed Endings," *Renaissance Quarterly* 37.2 (1984): 207–227; esp. 221–222.
32. Gallagher, "Oracle," 483.
33. Lori Newcomb, "'If that which is lost be not found': Monumental Bodies, Spectacular Bodies in *The Winter's Tale*," in *Ovid and the Renaissance Body*, ed. Goran V. Stanivukovic (Toronto: University of Toronto Press, 2007): 239–259; esp. 239–240.

[... Third Link ...]

Seeing the Spider: Cognitive Ecologies in *The Winter's Tale*

Gail Kern Paster

In Shakespeare's *Winter's Tale*, the clearest instance of an emotion being coproduced environmentally occurs early in the second act, when Leontes contrasts himself to the man who escapes being poisoned by not knowing of a spider's presence in the cup he has just emptied:

> but if one present
> Th' abhorred ingredient to his eye, make known
> How he hath drunk, he cracks his gorge, his sides,
> With violent hefts. I have drunk, and seen the spider. (2.1.44–47)[1]

The idea of a contingently poisonous spider was a proverbial one, presumably known to an early modern audience without benefit of scholarly gloss, although (as often the case with Shakespeare) the lines explain themselves well enough on their own. Through this trope, Leontes wants to insist that his experience is not the hypochondriacal self-poisoning of an over-imaginative man jealous by nature or, as Emilia describes such a one, "jealous for they're jealous" (*Othello* 3.4.158). He believes the nature of his torment to involve a different metaphorical scenario about an object with preternatural power to cause the cognitive emotions of horror and amazement.[2]

While we know that Leontes is disastrously wrong in believing himself betrayed, his trope of the spider in the cup represents a striking example of early modern cosmological thought. By constituting the spider as a thoroughly subjectified object (inside and outside the mind of the perceiver), by thoroughly dissolving the borders between the social and the biological, this trope affords a perfect instance of the ecology of the passions. In the proverbial usage, the spider is not simply toxic on its own, since its efficacy requires the victim's apperception and belief. For the victim, incorporation takes two separate forms—taking in the poisoned drink itself and taking in knowledge of it. But, thanks to an ambiguously inflected social transaction in which the "abhorred ingredient" and the fact of its being consumed are revealed to the victim, responsibility for the poisoning is powerfully shared not only between subject and object but also by an unnamed other. All three act together to turn abhorrence into poisoning and to bring about an unwilled dissolution of self through the body-opening violence of the involuntary purge. Being poisoned requires two kinds of external transactions, then, both natural and social—since in this scenario the drink's toxicity would not have been known except for the malice, or at least the superior knowledge, of the presenter and—just as important—it might not have existed except for a combination of appetites and emotions in the drinker: thirst, belief, fear, and abhorrence. The spider's power over the victim is simultaneously biological and psychological, actual and unreal, natural and social—an example of what Merleau-Ponty may have meant by writing that "a psychology is always brought face to face with the problem of the constitution of the world."[3] Because there is, after all, a third scenario, in which the presenter reveals the spider's presence *before*, not *after*, the cup has been drunk; the spider's abhorredness is then robbed of its potential to do bodily and psychic harm thanks not to ignorance but to an act of friendship possibly still visible to us if not to Leontes.

In a psychoanalytic reading of the play, the toxin here is the maternal body, mimicked by the abhorrent spider's swollen belly (implied rather than described). And the essential conflict in the opening action of the play occurs almost entirely within Leontes as he recoils from the spectacle of male loss represented by the psychically overwhelming plenitude of Hermione's pregnant body. For Janet Adelman, this intraphysic crisis instantiates a wider cultural anxiety about masculinity.[4] For David Hillman, Leontes is a version of the Shakespearean skeptic whose "refusal to acknowledge another is figured as a quelling or denial of the contents of one's body."[5] For both critics—however different their critical investments in the play—the essential action is internal to Leontes, and the cast of characters in this part of the play reduces to two. But, while acknowledging the force of such interpretations especially for us, I want nonetheless to offer another reading that may ring truer for an early modern understanding of the passions and their embodiment. In this reading, the imagined minidrama of the spider in the cup moves the action of paranoid jealousy out beyond the consciousness of the jealous person himself and embeds it in a much wider social and natural landscape, one where the spider, the presenter, and the victim coenable the act of violent disgust and purgation. Leontes's jealousy, in such a view, is constituted reciprocally by the world in which he lives and the world that lives in him.

For the early moderns, Leontes would necessarily live in the same kind of natural environment they did—one endowed with appetites that bound nature together in chains of sympathy and antipathy. In *The Anatomy of Melancholy*, Burton writes, "In vegetall creatures what soveraignty love hath by many pregnant proofes and familiar examples may be proved, especially palme trees, which are both he and she, & expresse not a sympathy but a love passion."[6] Francis Bacon writes of the desire for union to be found in various substances, from the strong desire of liquids for merging with other bodies to the strong resistance to such union in materials like iron or wood. And he describes bodies of a middle sort—"those bodies which are clammy and cleaving are such as have a more indifferent appetite (at once) to follow another body and to hold to themselves."[7] Even the behavior of minerals manifested occult desire, the shepherd Dorus explains in Sidney's *Arcadia,* comparing his own love to the "seemingly insensible loadstone, with a secret beauty holding the spirit of iron to it."[8]

We ought to regard these accounts of desire in the object world as part of the ecological framework of early modern personhood, which Timothy Reiss has usefully characterized with the Latinate term *passibility.* Passibility denotes "experiences of being whose common denominator was a sense of being *embedded in and acted on* by [circles or spheres]—including the material world and immediate biological, familial and social ambiences . . . and cosmic, spiritual or divine life." These circles, says Reiss, "*preceded* the person, which acted as *subjected to* forces working in complicated ways from 'outside.' But because of the embedding, that 'outside' was manifest in all aspects and elements of 'inside'—of *being* a person."[9]

The experience of seeing the spider in the cup, the experience that Leontes sees as a likeness of his own state of mind, is a crisis of the passible subject becoming aware of the intricacy of its embedding in a world whose material, social, and familial properties he must but can no longer take on faith.[10] In such a crisis the spider is not, psychoanalytically, the body of the mother alone but, ecologically, the body of the entire world—suddenly abhorrent, suddenly brought to conscious awareness by a malicious other, and violently regurgitated. It is the loss of this trusted world—the *oikumene* of home as world, world as home—which registers in Leontes's countenance "as he had lost some province, and a region/ Loved as he loves himself" (1.2.370–371).[11]

That resonant simile, uttered by Polixenes to Camillo when he first notices his "favour . . . begins to warp" (1.2.366), echoes the famously prelapsarian language of Polixenes's nostalgic account to Hermione of the kings' Edenic boyhood. But it also picks up the cosmological language used about the royal friendship at the very beginning of the play when Camillo and Archidamus work hard to construct the relationship not as contingent but as part of the order of things in which they all participate. It is organic, a rooted affection "which cannot choose but branch now" (1.1.24). It is a bond so strong that it collapses absence and presence, drawing the world into itself even as it projects itself onto the world. That which divides the two kings by culture and nature—that is, their adult responsibilities and the physical distances between their kingdoms—only serves to foster a rich stream of social exchanges in which royal messengers act as surrogate selves. Separation has moved Leontes and Polixenes to practice friendship as a form of global self-expansion, shaking hands "as over a vast" and embracing "from the ends of opposed winds" (1.1.29–31). The image hyperbolically projects the kings' embrace onto the globe, making their affection part of the map of the world. "The heavens continue their loves," Camillo declares (1.1.31–32), partly in order to invoke divine blessing but also to endow their friendship as a principle of cosmological order. Archidamus agrees: "I think there is not in the world either malice or matter to alter it" (1.1.33–34).

This is not the picture of an affection co-opting the world—as one might say about the love affair of Antony and Cleopatra—but of a world and a friendship involved in a course of reciprocal self-construction. Polixenes, too, sees the friendship as both self- and world-creating. His narcissistic image of the "twinned lambs that did frisk i'th' sun" (1.2.69), in its exclusive focus on the two boys, seems to me a childhood version of the courtiers' global imagery of royal self-extension. Even when he complicates this simplistic image by referring to the temptations since then "born to's," Polixenes maintains the cosmos's determining influence upon personhood by using the discourse of the passible self as it is precipitated out of mere animal consciousness by the combined workings of culture (arriving in the tempting form of wives) and of time and nature (arriving in the form of strong blood). Polixenes presents personhood as a kind of environmental affordance—the result of forces inside and outside the body working in tandem to produce high-spirited selves capable of desire, language, and morally complex behaviors such as marriage or sovereign sway. Being poisoned by the spider of perception works much the same way.

The cosmological terms of these two conversations—the royal friendship projected onto the globe and the image of twinned lambs metamorphosing into appetitive, sexually knowing men—offer a rich phenomenological backdrop to the onset of Leontes's jealousy and his image of the spider in the cup. While an account of that jealousy must always come to terms with Hermione and the psychological significance of her pregnancy, it should not do so by scanting the rich constitution of Leontes's world apart from Hermione. Even before he likens his sense of betrayal to the discovery of the spider in the cup, the discourse of Leontes's jealousy makes clear that the properties of the world are never innocent of human intention or consciousness. Leontes recoils from the appetite for union that he thinks he perceives in Hermione and Polixenes, declaring "to mingle friendship far is mingling bloods" (1.2.111). It is not surprising that he would look to the welkin-eyed Mamillius for reassurance about the nature of the external world, about the circles of their common embedding in and by Sicilia. The boy is variously a calf, an egg, a kernel, a squash, a gentleman, an honest friend. In his unbreeched, unfledged state, lacking his father's "rough pash" and "the shoots that I have" (1.2.130), he is an image not only of lost boyhood but, in his ongoing change and growth, an alternative image of the natural world to Hermione's suddenly abhorrent, suddenly spiderlike pregnancy. Yet this use of Mamillius keeps failing to reassure him because—as the image of the spider in the cup suggests—the external properties of the world change their nature with the attitude of the viewer. For Leontes, the spider in the cup, "th' abhorred ingredient" toxic to those who believe in it, is the phenomenal world itself constructed first for him by friendship but taking its mature form in the manifold circles of passibility represented by a kingdom, a wife, and a child. That he as perceiver is complicit in that world's sudden toxicity is part of the burden of this crucial trope, and suggests why cognitive and emotional acceptance of the world's affordances can be restored to Leontes only through the sight of Hermione's statue coming to life and the transformation of abhorrence into wonder.

[...]

NOTES

1. References to Shakespeare's plays are from *The Oxford Shakespeare: The Complete Works*, ed. John Jowett, William Montgomery, Gary Taylor, and Stanley Wells (Oxford: Oxford University Press, 2005).
2. On these cognitive emotions, see Lorraine Daston and Katharine Park, *Wonders and the Order of Nature 1150–1750* (New York: Zone Books, 1998), 15.
3. Maurice Merleau-Ponty, *The Phenomenology of Perception*, trans. Colin Smith (New Delhi: Shri Jainendra Press, 1962), 60.
4. Janet Adelman, *Suffocating Mothers: Fantasies of Maternal Origin in Shakespeare's Plays, Hamlet to the Tempest* (New York: Routledge, 1992).
5. David Hilllman, *Shakespeare's Entrails: Belief, Skepticism and the Interior of the Body* (London: Palgrave Macmillan, 2007), 153–154.
6. Robert Burton, *The Anatomy of Melancholy*, ed. Thomas C. Faulkner, Nicolas K. Kiessling, and Rhonda L. Blair, 6 vols. (Oxford: Clarendon Press, 1989–2000), 3:42.
7. Francis Bacon, "Experiment solitary touching appetite of union in bodies," in *Sylva Sylvarum: Philosophical Works, Vol II., The Works of Francis Bacon*, 7 vols., ed. James Spedding, Robert Leslie Ellis, and Douglas Denon Heath (London: Longmans, 1876), [Cent.III.293], 438.
8. This is a commonplace observation but here I quote Sir Philip Sidney, *The Countess of Pembroke's Arcadia,* ed. Victor Skretkowicz (Oxford: Clarendon Press, 1987), 165.
9. Timothy J. Reiss, *Mirages of the Self: Patterns of Personhood in Ancient and Early Modern Europe* (Stanford: Stanford University Press, 2003), 2.
10. See Robert N. Watson, *Back to Nature: The Green and the Real in the Late Renaissance* (Philadelphia: University of Pennsylvania Press, 2007), for an argument that links Leontes's skepticism as of a piece with Cartesian skepticism (51–66).
11. For this definition of the ancient Greek term, I am indebted to John Gillies, *Shakespeare and the Geography of Difference* (Cambridge: Cambridge University Press, 1994), 5.

8 "There's Magic in The Web of It"
Skin, Mind, and Webs of Touch in *Othello*

Jennifer Rae McDermott

> The mind is then moved, in its very *inspectio*, even into extension, and the latter is extended all the way to the mind, through all the channels and all the fibers of the body, where the mind exposes itself by inspecting. The soul is then touched: meaning at once that it is disturbed and that it is impressed by the "imprinted vestiges" in the body, in other words, by the extended traces of the world's extension.
>
> —Jean-Luc Nancy, *Corpus*

> 'Tis probable and palpable to thinking.
>
> —William Shakespeare, *Othello*

Rene Descartes's name has become synonymous with the separation of body from mind, and so it is all the more surprising and wonderful that Jean-Luc Nancy should choose Descartes's example of a man handling a piece of wax in the *Second Meditation* as a means to question how bodily extension and pure cogitation come to "*touch one another*."[1] Nancy asserts that in this instance "thought is sensing" and, he further specifies, a "sensing that still remains as touching."[2] The present essay undertakes a similar project: I use the sense of touch to map the intersections of body, mind, and affect found in the web metaphors from William Shakespeare's *Othello*. Ultimately, I argue, Shakespeare embodies jealousy in this play as a form of cognitive (and cutaneous) pressure that touches the audience. Where Nancy implies that the curious mind, *inspectio mentis*, can be touched, impressed, or moved by the external world via extensions within the physical body—its "channels" and "fibers"—he reprises an early modern language of thought rendered palpable in a web of touch. Early in *Othello*, when Brabantio pauses to consider how he lost his daughter, he remarks that his belief is "palpable to thinking" (1.2.76).[3] Even as his mind grapples with the thought and apprehends it, he orders the guards in turn to "apprehend" and "attach" Othello (1.2.77). The imperative charge to "hold" the criminal (1.2.79) continues the pun, since *holding* derives from the Old French *hend* and Old English *healdan*, giving rise to *ad + prehend- ĕre*. His jealous thought is thus "palpable" because it conflates webs of skin and mind into one felt conviction, crossing between manual apprehension—"the true office of the Hand is to apprehend or to holde"—and mental comprehension.[4]

But what is the real relationship between affective feeling and sensory feeling? Or, in other words, when does a feeling become touch? This slippery terrain becomes all the more difficult to negotiate when we return to the early modern definitions of touch as a sense. *To touch* was a verb used interchangeably with *to feel* in the early Tudor period, as Carla Mazzio reminds us, and both terms implied "physical as well as emotional sensation."[5] Thomas Cooper's 1578 definition provides a ready example of this interpenetration of passionate and sensate experiences since, for him, *to touche* meant equally "to mooue or grieue."[6] Richard Brathwaite likewise writes in his 1620 *Essaies Vpon the Five Senses* that "the Touches mouingst action" is "the *feeling* of his passion," and he includes the irascible and concupiscible among the unique "objects" of "this Sence of Touch."[7] The turn from somatic to affective touching in Brathwaite's essay hinges on the mind's susceptibility to being inwardly "moued" even as the body outwardly touches, and so too Shakespeare's *Othello* calls attention to the productive ambiguity of psychophysiological "feeling." When Iago winces at the close-felt pain of a verbal challenge saying, "Touch me not so near" (2.3.210), for instance, he means to show his supposed sensitivity as a friend through a metaphor of physical proximity even though he escapes true injury in the brawl. What Iago here perceives as a "touch" is not tactile pain in the purest sense, but rather the mental feeling of pain through emotion realized in sensory metaphor—his touch becomes feeling when *Othello* bodies forth the point of contact between palpation and cognition.

My consideration of Shakespeare's *Othello* centers on precisely this aspect: the overlapping webs of touch and thought that foster a particular "feeling" in the audience. Revisiting a pre-Cartesian understanding of an embodied mind through current cognitive theory, I hope to suggest why it is that psychologists should seize upon this early modern play to define feelings of extreme conjugal paranoia: dubbed the "Othello syndrome" by John Todd in 1955.[8] Of course, Shakespeare's nuanced presentation of spousal jealousy is oversimplified when used as a label in these later case studies, but they nevertheless remain faithful to the play's representation of this painful emotion in that they likewise reveal a crossover between touch perception and jealous cognition. Touch, even more immediately than the other senses, invites a rethinking of binaries such as body and mind, activity and passivity, agent and recipient. Contact is in its very nature dependent on reciprocity or, to invoke Merleau-Ponty's term *intercorporeity*: to touch is also to be touched.[9] As Eve

Sedgwick argues in her consideration of affect and texture in *Touching Feeling*, the sheer complexity of *touch* necessitates nondualistic thinking: "to touch is always already to reach out, to fondle, to heft, to tap, or to enfold, and always also to understand other people or natural forces as having effectually done so before oneself, if only in the making of the textured object."[10] Mazzio brings Sedgwick's "touchy-feely conception" into an analysis of early modern theatrical experience through "tact" in what she calls an "aesthetics of feeling."[11] Both Mazzio and Sedgwick emphasize the resistance of touch to definition, treating touch as a subset of the unspeakable or inarticulate.[12] By contrast, my reading of touch in *Othello* is grounded in the opposite belief that touch underpins language and the embodied metaphors that shape expressions of emotion. Exploring these linguistic underpinnings, I endeavor to extend a parallel aesthetics of feeling.

Linguists, philosophers, and literary critics within cognitive theory agree that metaphoric language develops out of the experience of living in a human body, and that key sensorimotor concepts of up and down, inside and outside, and movement toward a goal shape neural structures, patterning how we perceive and describe our world. According to recent "neural modeling" studies, "perceptual mechanisms and motor schemas" perform "conceptual work in language learning."[13] Crucially, the sense of touch not only uses webs to extend physical perception from body into the environment but further touch furnishes the web of cognitive metaphors we use to define that same environment. Mark Johnson notes in his integrated theory of primary metaphor that for a young child the "subjective experience of affection is typically correlated with the sensory experience of warmth" in touching, or "the warmth of being held."[14] This gives rise to linguistic forms later in life such as a "warm smile," a "close" acquaintance, or even an old "flame." Shakespeare's love of puns makes his drama ideally suited to explore the multiple meanings of words, as Mary Thomas Crane deftly illustrates in *Shakespeare's Brain*. Of all his plays, *Othello* is the most densely packed with touch metaphors that intertwine the holding of thoughts with the beholding skin as a way to convey "touching" affect to the audience.

Touch is the dominant sense in *Othello*. This claim may seem unsettling for a play so often thought about in visual terms of black and white. Yet *Othello* contains extensive references to characters touching, kissing, smothering, stabbing, caressing, or coming into contact with one another through various tactile means; and—while I primarily address webs and skin—hands (another early modern emblem of touch) appear virtually everywhere in this play. Hands and the metonymic handkerchief are invoked more than sixty times, more than in any other Shakespearean drama save *Titus Andronicus*, where literal dismemberment occurs. Indeed, physical contact abounds: Brabantio begrudgingly gives his blessing to Desdemona and Othello by joining their hands (1.3.190);

Cassio makes bold "show[s] of courtesy" with Emilia in kissing her and with Desdemona in "taking her by the palm" (2.1.100, 165); Othello grabs Iago "by the throat" (3.3.360), strikes his chest (4.1.179), and slaps his wife (4.1.234), to name but a few examples. What makes Shakespeare's punning on outward and "inward tuch" in *Othello* unique, I argue, is the way in which he combines these touch metaphors with imagery of spiders and webs appropriate to the Renaissance iconography of touch—at once overlying the fibrous web of skin, the neural web of ideas, and a social network within the theatre—in order to materialize jealousy as the felt pressure of weight and "[t]his heavy act with heavy heart relate" (5.2.376).[15]

SKIN AND SPIDERS

In Renaissance allegory and iconography, touching and arachnoid sensitivity are consistently linked. John Davies outlines this correspondence in his poem *Nosce teipsum* (1599), describing "the *Feeling Power*" in his catalogue of the five senses:

> Through every liuing part it selfe doth shed,
> By *sinewes* which extend from head to foote,
> And like a Net all ore the body spred.
> Much like a subtill Spider, which doth sit
> In the middle of her Web, which spreadeth wide,
> If ought do touch the vtmost threed of it,
> She feels it instantly on euery side.[16]

In this layered simile, touch is like both the "web" itself, spread across the body in the skin's "sinewes," and the "spider" that "instantly" feels any disturbance. Made to act as both the instrument and the means of deciphering stimuli delivered by that instrument, touch is uniquely self-reflexive. Through the web's extension of the body, the spider senses with remarkable range and nuance. Moreover, it not only "feels" but infers changes in the surround with a power homologous to the thinking brain. In Davies's words, "touch" is the "outward Instrument of sense" through which "euery thing must passe,/ Ere it approach the mind's intelligence,/ Or touch [our] wits."[17] The web symbolizes both touching and knowing because its outward-reaching tactile filaments allow the perceiver to gather "intelligence." Davies's spider-touch draws from a long philosophical tradition that likened the fibrous skin to a net and further used that simile to explain "intellectual apprehension"—a parallel Davies borrows from medieval depictions of tactile webs, which derive in turn from Aristotle's fourth-century treatise *De Anima*, which associates "exactness of discrimination" in "touching" with the keenness of man's "intelligence."[18] An image of both cutaneous and cognitive sensitivity, the web represents the powers of extension available to the spider through touch.

Touch is traditionally imagined as the only unmediated sense because it brings two bodies into direct contact without the intervening air or water necessary for other perceptions. For this reason, touch is *the* defining trope of contiguity. Yet as Davies's web analogy suggests, touch is not always bounded by the body that makes contact. Rather, prosthetic technologies of extension, such as the web, can artificially enlarge and stretch touch across the air into the environment, moving from the physically immediate to the relatively distant in a corresponding shift from sensory to "extrasensory" gathering.[19] The spider's web offers an exceptionally clear example of what I call "extramediated touch," although the same principle inheres in skin and other forms including (as I will show) Othello's handkerchief. The web is, in Freud's familiar words, an "auxiliary organ" or "artificial limb" because, while it is not quite "part of" man, it is nevertheless what "man has become."[20] Like the technologies of extension Freud lists in *Civilization and Its Discontents*—the microscope that improves vision and the telephone that allows the ear to hear across distances—extramediated organs of touch draw the external world nearer to the point of "touch[ing]" our "wits."[21] The spider's web, as an embodiment of both pure and extramediated touch, supplies Shakespeare with a provocative sensory image to test the limits of jealous sensitivity.

Early in the second act of *Othello*, there is a moment where Iago reveals to the audience how touch inspires his plot against the moor. Watching Cassio caress Desdemona in the ritualized gestures of a courtier's welcome, Iago whispers in his aside: "He takes her by the palm . . . / With as little a web as this will I ensnare as great a fly as Cassio" (2.1.166–167). The web of her skin, as "smooth as monumental alabaster" (5.2.5), carries the associations of erotic entanglement familiar from Edmund Spenser's allegory of seduction in the Bower of Bliss, where Acraisa's "alabaster skin" becomes the "subtile web" that enfolds her lovers.[22] Notably, while Caroline Spurgeon devotes just two pages to touch in her seminal study of Shakespeare's imagery, she observes that "there is one substance in especial to which he is markedly sensitive, and that is the texture of the skin."[23] Shakespeare's heightened awareness of skin is likely a consequence of his upbringing, because his father's trade involved tanning goat and deer hides. John Shakespeare was a "whittawer (that is, a specialist in the preparation of soft, white leather)," and Shakespeare may even have helped to prepare these skins before stitching them into gloves.[24] Iago focuses on the skin of the hands, mocking: "well kissed and excellent courtesy!" (2.1.173) and sneers in a bawdy reference: "Yet again, your fingers to your lips?/ Would they were clyster pipes for your sake!" (2.1.174–175). Comparing the fingers to pipes used in the late Elizabethan period to administer vaginal douches, Iago converts an innocent touch into a perversely penetrative gesture. Yet he does not plan to "strip [Cassio] out of lieutenantry" (2.1.171) by having him caught up in these flirtations in earnest. Rather, Iago uses the substance of Cassio's physical "tricks" with Othello's wife to spin a tale of adultery (2.1.170).

Iago is in many ways an ideal spider; early modern literature on "Spiders and their several sorts" aligned these "venomous creatures" with cunning intelligence and entrapment as well as touch.[25] Edward Topsell's description of the spiders who observe "very diligently the stirring of their deceitful webs, and perceiving them moving, though ever so lightly" applies to Iago as he threatens the great "fly" Cassio: "it had been better you had not kissed your three/ fingers so oft, which now again you are most apt to play/ the sir in" (2.1.171–173).[26] To Topsell, arachnid success depends on web sensitivity and deceptive ability. Iago's word *apt* emerges in this context in its original sense, meaning to fasten or bind (from Latin *apo*, "fitted"). As Joel Altman notes in his study of improbability, "*apt* does not purport to describe the thing as it is in itself but rather as it is in relation to other things."[27] Our mental assessments necessarily move between categories of probability in determining a course of action: if something is "apt," then it seems likelier to be true. Iago seizes upon this web of "apt" relations between suspicion and "other" apparently true "things."

Uniquely inclined toward logical supposition and an accompanying language of ligature, Iago is the only character to use the word *apt* in the play. In this respect, Iago troublingly resembles the playwright, because he not only recognizes plausible fictions but delights in the art of weaving lies and puppeteering other characters within a web of invisible "strings" (1.1.3). Iago creates conflict by using Cassio's "apt" intimacy (2.1.172) to develop a credible story—"she loves him 'tis apt" (2.1.281)—and remarks on Desdemona's murder: "I . . . told no more/ Than what he found himself was apt and true" (5.2.181). Just as Shakespeare brings his play together around Othello's paranoid sensitivity, so too Iago uses his sensitive touch to gather a web of text. The origin of *text*, as Elizabeth Harvey explains, is "from the Latin *tex-ere* (weaving)" that evokes a "sense of the literary work's tactility, its woven, web-like, or indeed, cutaneous properties."[28] When Iago advises Cassio to seek Desdemona's "apt" aid (2.3.309), her appeals become the web that substantiates his accusations:

> And by how so much she strives to do [Cassio] good,
> She shall undo her credit with the Moor.
> So will I turn her virtue into pitch,
> And out of her own goodness make the net,
> That shall enmesh them all. (2.3.345–349)

The sticky texture of "pitch" makes the inevitability of capture all the more frightening: "striving" to break free of the fibers that entangle them, Desdemona and Cassio pull the web only tighter by trying to escape. Iago's language of aptness in *Othello* offers a representation of entrapment based on the linkages between ideas within the mind that tie suggestion to inference or connect probable thought to its palpable conclusion. Iago creates these intangible ligatures even in the same instant that his metaphor of the net depends upon tactility to give the impression of jealousy or "her body's lust" physical form.

Iago's invocation of a net to "enmesh" the lovers clearly alludes to Vulcan's myth from book four of Ovid's *Metamorphoses.* As Marjorie Garber suggests, "*Othello* as a play is a kind of dramatic reworking of this tale of Mars, Venus, and Vulcan" where "Iago is the Vulcan figure, a tortured looker-on."[29] In the myth, the lame blacksmith god creates an invisible web of chains to catch his adulterous wife, Venus, in the act of coitus with Mars, and to hold them up to public ridicule: "he began, with subtle care, to fashion slender chains of bronze—so thin, the net and snare they formed could not be seen. . . . The slightest touch, the least of movements, was enough to set the web to work."[30] The wording invites comparison to Iago's soliloquy, not the least through its dual representation of the web as a means of capture and a means of perceiving the slightest movement. Where Iago's mind is "set to work" by observing the touches exchanged between Cassio and Desdemona, he appropriately plans a snare of touch to trap them. To speak of touch and tapestry also recalls the classical figure Arachne: the unfortunate maiden of Lydia who challenged Minerva to a weaving contest, lost, and became a spider. By making her sexual charges against the gods palpable in fabric, Arachne took her talent too far. While some myths explain her transformation as an act of mercy—Arachne hung herself but was restored by the goddess to continue her spinning—others, such as Spenser's intriguing "Muiopotmos" (1591), say that the maiden converts herself into a spider through "the sheer poison of envy."[31]

Othello's jealousy takes the form of a spider that poisons the mind once infected by Iago: the man who ironically warns him, "O, beware my lord, of jealousy!/ It is the green-eyed monster, which doth mock/ The meat it feeds on" (3.3.167–169). Spiders are notoriously cruel eaters, indefinitely holding their prey captive while "mocking" the meat they feed on. Topsell states that spiders not only delay in their consumption of trapped insects but further use touch to seduce the prey before it is devoured: "so soon as they espy their enemy to be caught in their nets, they do not first of all bite and prick him to death in any hostile manner, but they seem with their feet gently and softly to stroke him . . . untill they have throughly insnarled him within their clammy and viscous gins."[32] Whether or not Shakespeare envisaged the monster of jealousy as a spider, its green eyes offer a clue as to the way Othello's passions obscure vision to turn the body instead toward touch. The association of the color green with envy is often attributed to Shakespeare from Iago's phrase the "green-eyed monster," and Portia's similar invocation of "green-eyed jealousy" in the *Merchant of Venice*, but this relationship has much older origins. Galenic medicine believed that envy built up bile, which gave the skin a distinctive green hue. Further, the word *green* in Middle English was not merely the name of a quality but a verb: "it was an action you could do" or something you could perceive "*with* . . . 'to green' was 'to desire, to long.'"[33] Thomas Wright notes that individuals suffering from an "inordinate affection" have increased errors in judgment because they "seeth all things convenient most aptly" as if wearing "green spectacles, which make all things resemble the colour green."[34] If someone is already slightly jealous, then they are more prone to viewing any action in the light of their jealousy. This is what Iago relies on when he tells Othello: "Look to your wife; observe her well with Cassio;/ Wear your eyes thus" (3.3.199–200). His instructions impose Wright's spectacles, calling Othello's jealous mode of looking into effect paradoxically by claiming to safeguard against it. Yet *Othello*'s consistent integration of touching and knowing implicitly proposes touch as an alternative to the unreliability of "ocular proof" (3.3.361). When Othello commands, "Make me to see't, or at the least so prove it/ That the probation bear no hinge nor loop,/ To hang a doubt on," his elusive metaphors of "hinge" and "loop" reprise the tactility of weaving (3.3.365–367). Iago presses the issue of how much he must view to believe: "Would you, the supervisor, grossly gape on?/ Behold her topped?" (3.3.396–397). Recoiling from these thoughts, Othello instead demands "a living reason she's disloyal" or a tangible trace (3.3.410).

Of course the play's violent turn ultimately undoes the earlier promise of touch as corrective. Bound up with violence and sexuality, epitomized by Othello's suffocation of Desdemona, the surety touch provides is undercut by its own passionate "greenness." Spenser's allegory of the body uses a troop of "Snailes" and "spyders" to embody the fallibility of touch in eroticism's slimy traces: "With stings of carnall lust, and strong effort/ Of feeling pleasures."[35] When Iago repeats Cassio's sexual confessions in Act 4, Othello suffers an epileptic fit that resembles the effect of spider bites or "stings": "the patient is much disquieted . . . the tongue faltereth and stammereth," falling "down in great perplexity," and the "heart" is "turmoiled with an extraordinary kind of furious passion."[36] Othello embodies all these symptoms, even replicating the web through his speech pattern:

> Lie with her? Zounds, that's fulsome.—
> Handkerchief! confessions! handkerchief!—to confess
> and be hanged for his labour. First to be hanged,
> and then to confess: I tremble at it. . . .

> It is not words that shake me thus. Pish!
> Noses, ears and lips! is't possible? Confess? Handkerchief? (4.1.35–42)

His language weaves back and forth upon itself. The alternation of *handkerchief, confess, handkerchief,* and then *hanged/ confess* mimics the shuttle and bob of Arachne's loom. Interfolding the criminal erotic touch (linked to the tangible evidence of the handkerchief) and the punishment (hanging, like Arachne), Othello's language maps the network of his mind's suspicions in torn threads of logic. Envisaging what Iago suggests, Cassio lying "on her," Othello's mind collapses into pure "feeling" expressed bodily as trembling, shakes, and foaming at the mouth (4.1.54).

Near the end of the play, Othello avows that he is a man not easily made jealous, but one "wrought" to extreme perplexity (5.2.350). *Wrought* suggests that Othello's jealousy has been twisted into being by Iago's spiderlike activity. "Ensnar[ing]" his "soul and body" (5.2.307), Iago works upon Othello's mind the same way one might have "wrought" a textile through spinning or embellishing (*OED* 3.a.b)—a pointed example of which is the sibyl's embroidery on Desdemona's handkerchief. While the neural network analogy was first proposed by Alexander Bain in 1873 and William James, independently, in 1890, the language of "webs" has long permeated medical understanding. Reasoning that the brain must be the seat of the animal soul, Galen coined the term *rete mirabile*—or marvellous net—to characterize the system of blood vessels clustered at its base. This language of nets, webs, and weaving has not only endured but come to define the articulation of twentieth- and twenty-first-century cognitive science.[37] The physical extension of the body's touch into the handkerchief here continues as psychical extension in that Othello's brain is "wrought" like a web.

At the very heart of *Othello*'s problems of touch and sexual jealousy lies the handkerchief as a prosthetic piece of skin. Desdemona's handkerchief actualizes two early modern understandings of touch in one palpable object: it is both a somesthetic web and a concrete item metonymically connected to acutely sensitive hands.[38] Constantly passing from one character to another, stroked lovingly by Desdemona, snatched up by Emilia, grabbed by Iago, held by Cassio, Bianca, and then Cassio again, the handkerchief remains quite literally the matter at hand. Many scholars have read the handkerchief as a replacement for Desdemona's private body: an item of contiguity forming part of her domestic corpus or a physical "fetish."[39] Where psychoanalysts have interpreted the strawberry spots as representing the virgin's blood-stained sheets, or even her torn hymen, the handkerchief has typically been thought of as an extension of the female body.[40] The napkin can also be read, however, as a pliable stand-in for Othello. This makes all the more sense because Desdemona personifies the prop and "reserves it evermore" about her person to "kiss and talk to" in her husband's absence as a material, and distinctly tactile, object (3.3.298). If the handkerchief is a piece of the body contiguous to but still foreign from the whole, then can it also be taken as a prosthetic hand or woven skin graft?

Othello's curiously exotic explanation of the handkerchief's origins would seem to imply yes. The handkerchief is an embodied object: it is a carefully preserved, composite fabric made out of embalmed flesh. As Paul Yachnin notes, it is an item of "wonder" and not simply a "square of embroidered cloth in a nation whose primary industry was the production of textiles."[41] Othello describes his gift as having been sewn together from threads "dyed in mummy"—a powder reduced from dead bodies—and the red upon the fibres emerges as blood conserved "skilfully" from "maiden's hearts" (3.4.70–73). Where the texture of the handkerchief mimics the sinewy nature of skin in its threadlike properties and silken surface, it is also infused with blood from others' hearts to become a living membrane. Simultaneously a very ordinary, English-style bit of "frippery," and also an extraordinary object of sexual investment and somatic extension, the handkerchief works—borrowing the anatomist Helkiah Crooke's words on skin—to "knit up the whole body" and play "together" through touch.[42]

IM-PRESSING THE AUDIENCE

From the opening lines of *Othello*, punning on touch embodies the weight of the mind at work. The play obsessively returns to the weightiness of ideas, emotions, or cognitive burdens as a felt pressure, where for example Desdemona aims to "touch" Othello's love with a request "full of . . . difficult weight" (3.3.81–82), or where she recognizes that she must "support" a "heavy interim" in his "absence" (1.3.256–257), feeling her husband's departure as "a bodily weight on her."[43] The sensation of touch is essentially a measure of force. The "soft" impress of light touches, such as a caress or Othello's "kisses" (5.2.16, 106), are perceived by the Meissner's corpuscles. These discs are associated with sensory dendrites that lie close to the skin's surface, able to be bent with little force. Intense pressure such as the smothering murder of Desdemona is felt as weight by the Pacinian corpuscles deep within the skin. The explicit contrast of light and heavy pressure in the final act embodies Othello's vacillating passions, tender love and forceful jealousy, and further it suggests how metaphors of weight in *Othello* come to bear jealous affect that crosses into the audience's felt sensations of sympathy.

Perhaps the best example of deep pressure as impression, or more specifically, "oppression," appears in the opening scene. When Roderigo shouts up to Brabantio that Desdemona has been carried off in Othello's "gross clasps," he assures his listener that he would recant his "wrongs" if "this be known to you" (1.1.25). The *know* of Roderigo's "if it be known to you" puns on the connotations of knowing as cerebral awareness and to have "known" her in the Hebraic sense of sexual intercourse. Transferring this jealous thought to her father, Brabantio's mind perceives its weight: "Belief of it oppresses me already" (1.1.139). While Brabantio's expression equates mental difficulty with physical burdens as a familiar trope, one so common that it hardly registers as a metaphor at all, it is worth reconsidering because his phrasing grows out of embodied experience and language is "fundamentally" grounded in the body of touch.[44] The wording chosen here is precise; Shakespeare's reference to Brabantio's mental suffering invokes a tactile sensation of pressure that causes his condition to register more forcefully with the spectator.

As Charles Frey attests in his work on dramatic response, the metaphors that stimulate bodily experience, as "kinaesthetic and proprioceptive sensitivity," yield a deeper "feeling" of "empathy" or physical identification with the characters.[45] The touch metaphor of oppression generates both abstract cognitive and concrete physical effects that create what Brathwaite calls "fellow-feeling."[46] Considering "The Sense of Others Misery," Brathwaite describes how one person can vicariously "suffer" another's "infirmitie."[47] This "compassion" is more than metaphorically a "sense" because, he avers, "men can resolve into teares, through their sensible feeling of others griefs."[48] True fellow-feeling involves shared pain experienced by touch as "sensible"; the sympathetic observer comes to "bear" the "burden" of another's sorrow "in his mind."[49] Similarly, theories of primary metaphor in cognitive linguistics link the subjective judgment of difficulty to the sensory domain of cutaneous pressure and muscular exertion. The experience in forming memories of "the discomfort or disabling effect of lifting or carrying heavy objects" is equated with the internal sensation of struggling under a new mass of ideas.[50] Here the *target* (muscular difficulty) enacts the *source* (mental burdens). Or, in more literary language, the skin under pressure is the implied *tenor* of "oppression" where the *vehicle* of that force is the "brain." This feeling (the impress of thought) relates to bodily sensation, for, as Jacques Derrida writes in *On Touching*, the "corpus, inasmuch as it weighs" also "in a certain way thinks."[51]

It seems particularly appropriate to evaluate the weightiness of *Othello*'s dramatic impression in terms of touch for this reason. Theatre is aligned with a physical capacity to "move," "grip," or "touch" an audience. As Thomas Heywood writes in *An Apology for Actors* (1612), "well-spirited action . . . hath power to new mold the hearts of the spectators and fashion them."[52] To be sure, Elizabethan and Jacobean writers believed that drama had the formidable ability to sculpt the playgoer emotionally, intellectually, or even morally. This affecting "touch" was a point of consensus among both contemporary advocates of the theatre and its most outraged detractors. Plays were believed to exert a material influence that either "consciously or subliminally" remodeled "patterns of cognition in its audience."[53] Heywood's verb *mold* necessarily implies tactility. John Northbrooke, in his antitheatrical treatise on *Vaine Playes or Enterluds* (1577), similarly sees the theatre as a place where people are "fleshlye ledde": "All such spectacles and shewes . . . enter our heartes" to "touche vs, and conuerte vs thereby both from God and good workes."[54] Importantly, this pressure can be understood both as a somatic response—where the play might have "hurt" the curious viewer or wrung out tears—and as an emotional movement that stimulates judgment of plot and character.[55] In the seventeenth century the word *emotion* meant stirring in a physical sense; it only later acquired the connotation of mental disturbance. While *motion* is the term Shakespeare uses to convey the psychosomatic feeling of interior pressure, only slightly later authors employed the word *commotion*. In *Paradise Lost*, Milton portrays Adam's reaction to Eve as "Transported touch: here passion first I felt,/ Commotion strange."[56] This description of "internal churning," as Michael Schoenfeldt observes, is a cogent predecessor of "our own 'emotion'" which is linked to visceral touch.[57] The early modern term *commotion* is especially applicable to theatrical experience because it introduces the *cum* ("with"); commotion is being set in motion "by" or "with" someone else. This form of touching depends upon a web of relationships among people within a shared space, since bodies do not simply possess "weight" but must "weigh on" one another "and *against* one another" to share force and create movement.[58]

In the earliest firsthand account of *Othello* in performance (1610), Henry Jackson repeats the term *moved* twice to describe the fitness of the tragedy, admitting: "they moved us to tears."[59] From the perspective of an early modern theatrical subject, a successful production of *Othello* generated a "moving" response in the somatic body. The palpable arousal of pity in theatrical identification causes a shared swelling feeling, whereby "witnesses" experience "a distending force effecting an enlargement from within."[60] This tumescence aligns interoception, or touch-perception in the viscera, with a more vague affective fullness often thought to be released by crying. Wright considers this link between theatrical touch and tears when he defines passion as "a sensual motion" or a type of physical force.[61] Passion moves the animal spirits within the body that in turn press themselves upon the heart, blood, and brain.[62] An intrinsic part of the body, the "force" of "strong passions," can be used to persuade, to "marvellously allure," or to draw the assembly into sympathy—with "tears dropping down their cheeks"—by way of their sensitive souls.[63] He offers advice about how best to achieve this affecting style of oratory, counseling: "if we intend to imprint a passion in another it is requisite first it be stamped in our hearts: for through our voices, eyes, and gestures the world will pierce and thoroughly perceive how we are affected."[64] The tactile verbs *imprint, stamp,* and *pierce* reveal a sensory bias in this communication.[65] Most interestingly, the passion must be "stamped" into the actor in order to "pierce" the hearts of the observers, forcing a mirroring between the participants in the materialization of the experience. Crooke equally insists that social events like theatre create a deep "impression into our Sense" that "imprints" palpably upon the "mind."[66] Passionate response, these authors would seem to say, touches us because it is an internal pressure we feel as weight.

Granted, the theatre's ability to affect its audience was defined by early moderns in terms of tactile force and physical pressure, but how, to borrow Bruce Smith's phrasing, "can the kinesthesia of this touching moment be communicated to someone outside the fiction?" Or how can "my sense of touch" allow "me to project what I can feel with my body *here* onto his body *there*, and reversing direction, to take what I see happening to his body *there* and feel it in my body *here*?"[67] To broach these questions about theatrical identification, particularly identification through touch and fellow-feeling, inevitably widens a consideration of emotion from an individual body into a study of commotion among multiple bodies that weigh against one another.[68] One of the most vital components of theatrical experience, as Crooke notes in his explanation of why we are more "recreated" with "playes acted vpon the stage" than learning "out of written books," is the moving effect of "society."[69] A select population of theatregoers could afford to pay for seats stage-side where they might be close enough to touch the actors, but, by and large, those assembled to watch *Othello* would remain at a distance from the performer—no more than fifty feet away, but no nearer than the margins. Crowding in the theatres encouraged social touching among audience members: "at playes in London, you shall see such heauing and shoouing, such ytching and shouldring to sitte by women . . . such tickling, such toying, such manning . . . that it is a right Comedie to marke their behauiour."[70] Philip Stubbes similarly depicts playgoing as a touch-based experience: "for what clipping, what culling, what kissing and bussing, what smouching & slabbering on another, what filthie groping and vncleane handling is not practiced?"[71] Leaving aside the jostling in the original Globe theatre performances, touch and the intersubjective interaction of people in close proximity may have had an additional influence on playgoers then as now. In face-to-face exchanges between people in society, studies have shown that one subject's sensorimotor coupling and emotional bodily signals can be replicated as force in another subject's sensorial brain to alter their felt response. By seeing and physically witnessing an event as felt by other "players in the web of biological processes," we respond by shifting our own physical state and reactions to that event.[72]

It is not anachronistic to envisage the relations among people within the theatre space as "players" in a web or "social network," because early modern theatregoers conceived of themselves as connected by invisible threads. "Sit in a full theatre," Thomas Overbury writes in 1616, "and you will thinke you see so many lines drawen from the circumference of so many eares, whiles the Actor is the Center."[73] The actor may reside at the heart of the circular web of touch, but his feelings radiate outward through the "lines" that extend to every member. Thomas Dekker's prologue likewise promises to hold his audience's attention by tying them to his body "with golden chains."[74] Northbrooke calls his reader to consider how "we kepe ioly cheare one with an other in banquetting, surfeting and dronkennesse . . . with Mummeries and Maskes" and exhorts them to give up these practices by referring to the fragility of our life within this shared "weauers web."[75] The physical interaction of bodies within the theatre venue materializes as a web of touch, not only through displaced contact but further as the social "net" of relations reduces the affective space between man and man in the sense of another's misery. Like the spider iconography of skin as a web that mediates touch throughout the body, the theatre becomes a fibrous network that ties playgoers together through the felt force of sympathy. By tracing the interpenetration of physical touching and cognitive reasoning throughout *Othello* in a combined web of "fellow-feeling," we can better understand the play's final dramatic effect: how it has moved our minds, impressed itself upon our bodies, and touched us with its weight.

NOTES

I am grateful for the insightful responses of Elizabeth D. Harvey, Jill Levenson, and Jeremy Lopez to earlier versions of this paper. This project was enriched by discussion with seminar members at the 2011 *International Shakespeare Association* World Congress, and my attendance there was funded in part by the generosity of *The Shakespeare Birthplace Trust.*

1. Jean-Luc Nancy, *Corpus*, trans. Richard A. Rand (Fordham, New York: Fordham University Press, 2008), 130.
2. Nancy, 131.
3. All citations of William Shakespeare's *Othello* are to the *Norton Critical Edition of The Tragedy of Othello, The Moor of Venice*, ed. Edward Pechter (New York: Norton, 2004).
4. Helkiah Crooke, *Microcosmographia: A Description of the Body of Man Together with the Controversies Thereto Belonging* (London: Early English Books Online, 1615), 730.
5. Carla Mazzio, "Feeling Inarticulate: On Communal Vulnerability and the Sense of Touch in *Lingua* and *Hamlet*," in *The Inarticulate Renaissance: Language Trouble in an Age of Eloquence* (Philadelphia: University of Pennsylvania Press, 2009): 175–220; esp. 180.
6. Thomas Cooper, *Thesaurus Linguae Romanae et Britannicae* (London, 1578), "*tango*, sv."
7. Richard Brathwaite, *Essaies Vpon the Five Senses with a Pithie Vpon Detraction* (London, 1620), A2, 31–32.
8. See Mary Thomas Crane, *Shakespeare's Brain: Reading with Cognitive Theory* (Princeton, New Jersey: Princeton University Press, 2000), 4. As she explains,

"cognitive science" investigates "the ways in which the mind (the conscious and unconscious mental experiences of perception, thought, and language)" is produced by the "brain and bodily systems." See also John Todd and Kenneth Dewhurst, "The Othello Syndrome: A Study in the Pyschopathology of Sexual Jealousy," in *The Journal of Nervous and Mental Disorder* 122 (1955): 367–374.

9. Maurice Merleau-Ponty, *The Visible and the Invisible*, ed. Claude Lefort and trans. Alphonso Lingis (Evanston, Illinois: Northwestern University Press, 1968), 141.
10. Eve Kosofsky Sedgwick, *Touching Feeling: Affect, Pedagogy, Performativity* (Durham, North Carolina: Duke University Press, 2003), 14.
11. Mazzio, 180–181.
12. See Didier Anzieu, *The Skin Ego/ Moi-Peau*, trans. Chris Turner (New Haven, Connecticut: Yale University Press, 1989), 13. He notes that "touch" has one of the longest entries in the *Oxford English Dictionary.*
13. George Lakoff and Mark Johnson, *Philosophy in the Flesh: The Embodied Mind and Its Challenge to Western Thought* (New York: Basic Books, 1999), 38.
14. Lakoff and Johnson, 46.
15. Philip Sidney, "Astrophil and Stella," in *Sir Philip Sidney: The Major Works*, ed. Katherine Duncan-Jones (Oxford: Oxford Classics, 1989), Sonnet 15.
16. John Davies, *Nosce teipsum* (London: Early English Books Online, 1599), 45.
17. Davies, 46.
18. Aristotle, *De Anima* (*On the Soul*), trans. J. A. Smith (Cambridge, Massachusetts: MIT Internet Classics Edition, 2009), 1.4.5 and 2.9.1. For more on the medieval history of touch and its visual representations, including the famous Longthorpe tower wall painting, see Sharon Assaf, "The Ambivalence of the Sense of Touch in Early Modern Prints," *Renaissance and Reformation* 29.1 (2005): 75–98.
19. Mazzio, 195.
20. Sigmund Freud, *Civilization and Its Discontents*, trans. David McLintock (London: Penguin, 2004), 36.
21. Davies, 46.
22. Edmund Spenser, *The Faerie Queene*, ed. A.C. Hamilton, (London: Pearson Longman, 2001), 2.12.77. On the connection between Acrasia and skin see Elizabeth D. Harvey, "The Touching Organ: Allegory, Anatomy, and the Renaissance Skin Envelope," in *Sensible Flesh: On Touch in Early Modern Culture*, ed. Elizabeth D. Harvey (Philadelphia: University of Pennsylvania Press, 2003): 81–102; esp. 87.
23. Caroline Spurgeon, *Shakespeare's Imagery and What It Tells Us* (Cambridge: Cambridge University Press, 1935), 82.
24. Ann Thompson and Neil Taylor, "Introduction," *Hamlet* (London: Arden, 2006): 1–137; esp. 37.
25. Edward Topsell, *The history of four-footed beasts and serpents* [. . .] *whereunto is now added, The theatre of insects, or, Lesser living creatures* (London: Early English Books Online, 1658), 769.
26. Topsell, 770.
27. Joel B. Altman, *The Improbability of Othello: Rhetorical Anthropology and Shakespearean Selfhood* (Chicago: University of Chicago Press, 2010), 90.
28. Elizabeth D. Harvey, "Introduction," in *Sensible Flesh: On Touch in Early Modern Culture*, ed. Elizabeth D. Harvey (Philadelphia: University of Pennsylvania Press, 2003): 1–21; esp. 14.
29. Marjorie Garber, *Shakespeare After All* (New York: Pantheon Books, 2004), 589.

30. Ovid, *The Metamorphoses of Ovid*, trans. Allen Mandelbaum (New York: Harcourt Brace), 116.
31. Judith Dundas, "Muiopotmos: A World of Art," *The Yearbook of English Studies* 5 (1975): 30–38; esp. 34.
32. Topsell, 783.
33. Bruce Smith, *The Key of Green: Passion and Perception in Renaissance Culture* (Chicago: University of Chicago Press, 2009), 37.
34. Thomas Wright, *The Passions of the Mind in General*, ed. William Webster Newbold (New York: Garland, 1986), 126.
35. Spenser, 2.11.13.
36. Topsell, 772.
37. See, for example, Richard L. Gregory, "Perception," in *The Oxford Guide to the Mind*, ed. Geoffrey Underwood (Oxford: Oxford University Press, 2001); 15–29; esp. 27, where "sensory experience" is linked to the brain by "tenuous threads of nerve." See also Michael Posner, *Cognition: An Introduction* (Glenview, Illinois: Scott, 1973). Posner regards memory formation as "active and continuously changing, like the weaver's loom" (2).
38. *Somesthetic* is the medical term for perception related to bodily sensations, specifically those of the skin, like temperature, pressure, and pain.
39. Harry Berger, "Impertinent Trifling: Desdemona's Handkerchief," *Shakespeare Quarterly* 47.3 (1996): 235–250; esp. 244.
40. On the handkerchief as synonymous with wedding sheets and its significance as a blood pledge, see Lynda E. Boose, "Othello's Handkerchief: The Recognizance and Pledge of Love," *English Literary Renaissance* 5 (1975): 360–374.
41. Paul Yachnin, "Magical Properties: Vision, Possession, and Wonder in *Othello*," *Theatre Journal* 48.2 (1996): 197–208; esp. 202.
42. Natasha Korda, *Shakespeare's Domestic Economies: Gender and Property in Early Modern England* (Philadelphia: University of Pennsylvania Press, 2002), 125; Crooke, 71.
43. Pechter, 22n.256.
44. Elizabeth Hart, "Cognitive Linguistics: The Experiential Dynamics of Metaphor," *Mosaic: A Journal for the Interdisciplinary Study of Literature* 28.1 (1995): 1–12; esp. 5.
45. Charles Frey, *Making Sense of Shakespeare* (London: Fairleigh Dickinson University Press, 1999), 32.
46. Brathwaite, 125.
47. Brathwaite, 124.
48. Brathwaite, 133.
49. Brathwaite, 125.
50. Lakoff and Johnson, 50.
51. Jacques Derrida, *On Touching—Jean-Luc Nancy*, trans. Christine Irizarry (Stanford, California: Stanford University Press, 2005), 71.
52. Thomas Heywood, *An Apology for Actors* (New York: Scholar's Facsimiles and Reprints, 1941), B4.
53. Louis Montrose, "The Purpose of Playing: Reflections on a Shakespearean Anthropology," *Helios* 7.2 (1980): 51–74; esp. 51.
54. John Northbrooke, *A treatise wherein dicing, dauncing, vaine playes or enterluds with other idle pastimes [. . .]* (London: Early English Books Online, 1577), 67 and 62.
55. Northbrooke, 68.
56. John Milton, *The Riverside Milton*, ed. Roy Flannagan (Boston: Houghton Mifflin, 1998), 8.525–26.

57. Michael Schoenfeldt, "'Commotion Strange': Passion in *Paradise Lost*," in *Reading the Early Modern Passions: Essays in the Cultural History of Emotions*, ed. Gail Kern Paster, Katherine Rowe, and Mary Floyd-Wilson (Philadelphia: University of Pennsylvania Press, 2004): 43–67; esp. 45.
58. Nancy, 93.
59. Jackson, cited in Yachnin, 205.
60. Robert Weimann and Douglas Bruster, *Shakespeare and the Power of Performance: Stage and Page in the Elizabethan Theatre* (Cambridge: Cambridge University Press, 2008), 22.
61. Wright, 8.
62. On Wright's passions and their role in understanding tragic emotion, specifically in relation to Shakespeare's *Macbeth* see Katherine Rowe, "Minds in Company: Shakespearean Tragic Emotion," in *A Companion to Shakespeare's Works,* Vol. I, ed. Richard Dutton and Jean Elizabeth Howard (New York: Blackwell, 2003): 47–72.
63. Wright, 90.
64. Wright, 212.
65. This stamping recalls the etymological root of the word *character* in that it signifies both a "distinctive mark impressed" or "engraved" (*OED* 1.a) and "the personality or 'part' assumed by an actor on the stage" (*OED* 17.a).
66. Crooke, 698.
67. Bruce Smith, "Touching Moments," in *Phenomenal Shakespeare* (Oxford: Blackwell, 2010): 132–176; esp. 134, 147–149, 151. Smith uses virtual reality studies to explain the crossover between feeling bodies in proprioceptive drift where subjects "react viscerally in their physical body when an object (in the experiment, a hammer) seemed to threaten their virtual body" or the body of another "dummy," aligning their sensory perceptions of touch with the touch of an exterior other.
68. Associated with these emotional experiences in mind-body activity are touch-related impressions, occurring both on the deep interoceptive level and on the surface exteroceptive level of the skin. For example, Neil Genzlinger remarked in a *New York Times* review of *Othello* from June 19, 2004, that the final scene never fails to "bring out a few goosebumps." This concatenated relationship between visceral arousal, emotional response, and cognitive evaluation in *Othello* relies on touch because, in many ways, "our feeling of bodily changes as they occur is the emotion": see William James, "What Is an Emotion?" *Mind* 9 (1884): 188–205; esp. 191. See also Giovanna Colombetti and Evan Thomson, "The Feeling Body: Toward an Enactive Approach to Emotion," in *Developmental Perspectives on Embodiment and Consciousness*, ed. Willis F. Overton, Ulrich Mueller, and Judith Newman (Mahwah: New Jersey: Lawrence Erlbaum, 2007): 45–68; esp. 58. The degree of touching we witness onstage also "stimulates our arousal state," which in turn produces an inner to outer feedback system that can affect "judgment thresholds" and "attention deployment." Seeing the characters touch, and feeling touched by them as we watch in the assembly, ultimately changes the way we think about them as characters.
69. Crooke, 698.
70. Gosson, Stephen, *The Schoole of Abuse*, in *Early Treatises on the Stage* (London: Shakespeare Society, 1853), 25.
71. Phillip Stubbes, *Anatomy of the Abuses in England*, ed. Frederick J. Furnivall (London: The New Shakespeare Society, 1877), 22.
72. Colombetti and Thomson, 62.

73. Thomas Overbury, "*The Conceited Newes*" *of Sir Thomas Overbury and His Friends: A Facsimile Reproduction of the Ninth Impression of 1616*, ed. James E. Savage (Gainesville, Florida: Scholars Facsimiles and Reprints, 1968), 209.
74. Thomas Dekker, *If It Be Not Good, The Devil Is In It. The Dramatic Works of Thomas Dekker*, ed. Richard Herne Shepherd (London: J. Pearson, 1873), 264.
75. Northbrooke, aiii.

Figure 4 "A Melancholic Encounter," from Johannes Curio, *Conservandae bonae valetudinis praecepta longe saluberrima* (1582), 235. Image A012388, History of Medicine Collection, Courtesy U.S. National Library of Medicine.

9 Coriolanus's Blush

Tiffany Hoffman

Making a display of his wounds in the marketplace is a part Coriolanus proclaims he "shall blush in acting" (2.2.142).[1] Why does Shakespeare's hero feel shame over his wounds and their exposure? Gail Kern Paster argues that the wounds threaten Coriolanus's heroic male agency through their association with the involuntary bleeding of the woman's body. Displaying his wounds would, as Paster suggests, implicate Coriolanus in a scene of "shameful feminization."[2] For Ewan Fernie, the exposure of the wounds would undermine Coriolanus's heroic self-image by drawing attention to his "fleshly vulnerability" while also reducing his perceived social status. As Fernie claims, in Coriolanus "there is a powerful class shame, of an aristocrat degraded below the common people."[3] Whether grounded in notions of rank, status, or gender, current work on *Coriolanus* has tended to theorize the hero's shame as an individualistic and nonmoral emotional response aroused through the loss of personal identity. In this essay I explore an alternative approach to shame by turning to Aristotle's account of the emotion as a fear of social dishonor. Aristotelian shame and its proximity to other affective phenomena—including shamefastness, modesty, and shyness—offer a new way of thinking about Coriolanus's blush as a social and ethical reaction that is deeply informed by the values governing the Republic and its code of civic duty in war. The great hero, I argue, fears that his blush will be read as a sign of moral transgression, disclosing the true meaning of his wounds as tokens of pride and heroic self-interest in battle.

As part of my investigation into the moral nature of Coriolanus's blush, I offer an extended historical account of Aristotelian shame, focusing on its translation into early modernity as a religious affect linked to conscience and prominently involved in theological attempts to address the element of bodily determinism inherent in Galenic theories of temperament. In accordance with this historical approach, the blush may be further conceived of as a psychophysiological response that moderates the hero's pride and calls his choleric nature into question. Elicited through the dynamic shifting reciprocities between self and environment characterizing humoral selfhood and emotional experience, Coriolanus's blush, I argue, motivates a process of moral consciousness and complexional reform through which his soul is purified and delimited as governing influence over his body. By mapping a Christian moral and affective framework onto an ancient Roman worldview, Shakespeare exposes his culture's tendency to regard "hard-line" accounts of Galenic humoralism and the embodied mind as irreligious.[4]

Galenism assumed that "the manners of the soul follow the temperature of the body."[5] The four humoral fluids—choler, melancholy, blood, and phlegm—traveled through the bloodstream to the heart, the seat of cognition and motion, where they were rarified into spirits that moved through the arteries to the brain, connecting the body's flesh and its immaterial soul. In *Coriolanus*, Menenius draws attention to this organic system as he explains to the Plebs how humors are sent "through the rivers of your blood/ Even to the court, the heart, to th'seat of the brain" (1.1.131–132). In the pre-Cartesian Galenic economy, mental processes were grounded in a dynamic balance of physiological processes. Humoralism, as Gail Kern Paster notes, "accounted for a person's thoughts and deeds in a way that did not distinguish, as we tend to do, between the psychological and the physiological."[6]

As derivatives of the bodily humors, the passions were understood as material forces that were felt as fluctuations in the heart. When the reason "sees occasion for emotion . . . the heart responds as directed by expansion or contraction. Coincidentally there is a movement toward the heart of the humor whose qualities will stimulate the proper reaction."[7] Experienced as humoral shifts within the body, the passions were thus capable of altering the body's complexion. As Lawrence Babb notes, if a passion is very strong, it can "establish the humor of corresponding qualities as the dominant humor . . . making the passion which produced it habitual."[8] In her seminal work *Humoring the Body*, Paster expounds upon the embodied emotions and their capacity for change, which she suggests can be properly understood only through a belief in the way the early modern body, mind, and world were always connected. Referring to a "premodern ecology of the passions," Paster observes how the quality and constitution of the air one breathed or the food one took in could shift the balance of humors in the body.[9] As Thomas Wright notes, "the humors depend upon the heavens, air, sleep and waking, meat and drink, exercise and rest," and "according to the alterations of these external causes one or other humor doth more or less over-rule the body, and so cause alteration of passions."[10] Yet, as John Sutton observes, within "these old physiological systems it was also assumed that the cultural environment was as influential a part of such conspiracies of causes as was the physical world."[11] Thus, just as temperament influences the passions, so the passions—when influenced by a myriad of social, natural, and external stimuli—may affect the temperament and even cause a permanent change in complexion.

The body's physiological openness to its environment meant that humoral subjectivity was characterized by a high degree of emotional instability. When impacted by forces in the outside world, the humors could rise to an unhealthy excess, causing illness and overtaking reason. Consequently the passions were often explained as part of the sensitive soul that humans shared with animals and regarded as unstable, irrational perturbations that corrupted "the judgment and seduc[ed] the will, inducing for the most part to vice, and commonly withdrawing from virtue."[12] Rational discipline equated with the Aristotelian mean was strongly encouraged as a means of attaining a moderate level of emotion, but it was not the sole form of control recommended.[13] Focusing on the humoral body's permeability with the physical environment, Michael Schoenfeldt recently investigated the role of food in maintaining health and well-being. Through the willing adoption of certain strategies of self-discipline, such as diet and exercise, it was possible to modify the body's "fluid economies for the desired physiological, psychological, and ethical outcome."[14] If, as Schoenfeldt clarifies, the belief in humoral temperament "appears at once deeply materialist and incorrigibly determinist, in actual practice it was possible to manipulate the humoral fluids and their concomitant behaviors."[15] John Sutton likewise explains how one's humors could shift with age and in response to a variety of factors. As he states, "biological temperament was just the dynamic mixture of fluids in different proportions and conditions, changing over time in accordance with external influences as well as the drying or cooling rhythms of the life-cycle. . . . Excessive or insufficient environmental or psychological input was a framework for explaining not only disease but also the varieties of health."[16] The humoral body was therefore not a rigidly fixed entity physiologically inclined to certain emotional and mental states. Rather, Galenism depicts emotional experience and human nature in perpetual flux, driven by the shifting balances in the body's fluids all in constant interaction with the world around it.

Coriolanus cannot be fully understood apart from humoral notions of subjectivity and material forms of emotional experience that inform his characterization. In his opening diatribe, the great hero refuses the people's request for corn, insults them through the election of their tribunes, and degrades their power in the Republic. Shakespeare portrays Coriolanus as a man "so choleric and impatient that he would yield to no living creature."[17] Through his passionate invective the hero is made into an antisocial force whose complexion engenders the uninhibited expression of pride and anger. Sicinius observes, "was ever man so proud as is this Martius?" To which Brutus responds, "marked you his lip and eyes?" (1.1.250–253). In the humoral economy governing Rome, physiognomic signs denote essences, the body is the "transparent signifier of one's fundamental being."[18] The tribunes clearly read Coriolanus's downcast gaze as evidence of his "surly nature," but their joking that such a contemptible expression "bemock[s] the modest moon" is ironic, since modesty is a quality that is not as inimical to the hero as is first made out to be (1.1.255).

According to Coppelia Kahn, Volumnia plays a "powerful part in framing her son's temperament and value system."[19] She is preoccupied with Martius's success as a warrior and hones his choleric passions of pride and anger from an early age so that she can reap the benefits, consequently nurturing within him an "essentialist consciousness."[20] Considering how "honor would become such a person," Volumnia gloats about the time she sent her "tender-bodied" son into a "cruel war" (1.3.6–13). Volumnia valorizes wounds as signs of heroic honor. Blood, she proclaims, more becomes a man than a gilded trophy:

> The breasts of Hecuba
> When she did suckle Hector looked not lovelier
> Than Hector's forehead when it spit forth blood
> At Grecian sword, condemning. (1.3.39–43)

Like Hector, Coriolanus sucks a kind of "valiantness" from his mother's breast (3.2.131). Volumnia's milk is converted into a liquid force that animates the hero's body, rousing his fury and driving him to pursue violent feats of "deed-achieving honor" (2.1.170). In Coriolanus, parental dependence is sublimated into unrestrained aggression but not completely overcome. "Framed by maternal insufficiency," Janet Adelman observes that, "Coriolanus can never wean himself from what he has never truly had: thrusting him out, Volumnia binds him to her."[21]

Although Volumnia declares that she would rather have eleven sons "die nobly for their country than one voluptuously surfeit out of action," her avowal that she was "pleased to let [Martius] seek danger where he was like to find fame" complicates the Republic's code of civic duty in war (1.3.24–25, 12–13). Volumnia's idea of *virtus*, which should encompass the integrated values of "love of blood and war, love of country, desire for a brave death and posthumous glory," appears self-serving.[22] As one citizen explains, "though soft-conscienced men can be content to say it was for his country," Martius acts to "please his mother and to be partly proud—which he is even to the altitude of his virtue" (1.1.34–37). Coriolanus may exhibit a superior degree of valor; however, Rome's "chiefest virtue" is morally deprecated to the extent that it is unpatriotically exhausted in battle in the interests of his mother's desire for glory, or his own, or both (2.2.82). The moral quality of Coriolanus's heroism thus remains in doubt. Nevertheless, it is difficult to ignore "what services he has done for his country," so the citizens excuse his martial conduct with recourse to the drives of the Galenic body, deeming his pride a "vice" he simply "cannot help in his nature" (1.1.27–28, 39). In their effort to exonerate Martius's pride, the citizens endorse the notion of physiological determinism inherent in Galenic theories of temperament. Despite the body's susceptibility to variability and discipline, humoralism left open the possibility for an individual to acquit himself of sin by claiming that "a person is wholly formed by his or her complexion," or that one's bodily fluids were involuntarily influenced by external factors.[23] Paster identifies the tendency for the early moderns to cite "their humors profligately to excuse lapses in behavior" as they invoked the undeniable "variety and obduracy of the physical body's appetites and their resistance to reason."[24]

As Douglas Trevor observes, this kind of "humoral scapegoating" was heretical. Explaining sinful conduct as an effect of the body and its fluids rather than the failure of rational self-control undermined religious notions of free will and "threatened the divine omnipotence of an electing god."[25] Such deterministically hard-line and inflexible accounts of Galenic selfhood also contributed to the limiting and irreligious belief that a particularly "severe form of predestination is manifested in the body," thereby thwarting the soul's journey toward immortality and rendering its corruption materially inescapable.[26] If the citizens pursue materialist explanations to excuse Coriolanus's unethical yet highly advantageous conduct in war, Martius himself appears to validate their thinking. Having been raised with an essentialist consciousness and with the belief that his choleric passions are vital to his heroic success, Coriolanus fervently strives to maintain a "constant temper" (5.2.92). Fearful that his complexion will shift through dietary means and that he will have a "suppler soul" after he has filled his "pipes and these conveyances of our blood with food and drink," Coriolanus practices "self-starvation" and "persistently regards food as poisonous" (5.1.53–55).[27] When Lartius warns him that his "exercise hath been too violent for a second course of fight," the hero ignores him and returns to the battlefield. Extended engagement in violent "exercise" offers Coriolanus a means of regenerating his choleric humors or, as Cominius puts it, "requicken[ing] what in flesh was fatigate" (2.2.115).

In view of the early modern formulation of emotions as bodily states that could become immoderate if not governed, overtaking reason and will, Coriolanus's pride progresses in war into the Aristotelian extreme of vainglory. As Aristotle maintains, "with regard to the desire for honor the mean is proper pride, the excess is empty vanity, and the deficiency is undue humility."[28] The tribunes discern how Martius accedes to the natural increase of his passions, noting how "the present wars devour him. He is grown too proud to be so valiant" (1.1.256). Martius is "overcome with pride, ambitious past all thinking" (4.6.33–34). During war nothing exists beyond the hero's pursuit of Aufidius, "a lion" he "is proud to hunt" (232–233). Proclaiming Aufidius to be "the man of [his] soul's hate," Martius discloses his perception of his soul as an entity influenced by bodily fluids that determine how he thinks, feels, and acts (1.6.10). In combat his disdain for Aufidius materializes into a liquid humoral force that makes him "sweat with wrath" (1.4.27). While searching for his enemy, the hero witnesses a man taken prisoner, powerless over the force of his passions; however, Martius demonstrates little control over himself and later regrets that he was unable to help him as "wrath overwhelmed [his] pity" (1.10.86).

Driven forward by bodily forces beyond his reasonable control, Coriolanus is envisaged as an irrational and unrestrained animal akin to a "fawning greyhound" that has "slipped" his "leash" (1.7.38–39).

Coriolanus's immoderate passions impel his vainglorious quest for his enemy, so that "alone" he enters the "mortal gate of the city" (2.2.108–109). The selfish aspect of Coriolanus's heroism is evinced after his single-handed attack on the Volsces, when he flaunts his wounds and invites approbation from the Roman army:

> [If any] love this painting
> Wherein you see me smeared; if any fear
> Lesser his person than an ill report;
> If any think brave death outweighs bad life,
> And that his country's dearer than himself . . .
> Wave thus to express his disposition
> And follow Martius. (1.8.69–76)

Coriolanus understands that *virtus* is a heroic attribute that invests martial deeds with social value. The fact that he assigns meaning to his wounds through the patriotic language of Roman *virtus* only reinforces how inordinate his passions have become, causing a "defect of judgment" to the extent that he is unable to perceive the immorality of his actions (4.7.39). Martius's pride, even if he appears unaware of it, ultimately breaks through as his men hoist him up in commendation and he vainly shouts, "O me alone! Make you a sword of me" (1.8.77). When he finally confronts Aufidius, he continues to boast of his singular defeat, proclaiming, "alone I fought in your Corioles' walls. . . . Tis not my blood/ Wherein thou seest me masked" (1.10.8–10). There is nothing Aufidius "abhor[s]/ More than thy fame and envy" (1.9.45). Disclosing the true meaning of the hero's bleeding body, Aufidius enviously calls attention to Coriolanus's wounds as tokens of vainglory, aptly demarcating the fame the hero has won through his sole defeat of the enemy.

During war Coriolanus courts wounds and engages in situations where he can openly display them as laudable signs of his heroic preeminence. Off the battlefield, however, any kind of praise is detrimental to the wounds, which "smart/ to hear themselves remembered" (1.10.27–28). To Cominius, Martius is simply being humble. The hero's derisive attitude toward praise "would seem but modest" (25). Yet the psychophysiological nature of Coriolanus's response—"I will go wash,/ And when my face is fair you shall perceive whether I *blush* or no"—reveals the hero's modesty to be a more complex affective state that gains ethical significance when read alongside Aristotelian accounts of the fear of shame (69; emphasis mine). Aristotle defines shame as a "kind of fear of dishonor," and suggests that it is morally operative. Because shame or "modesty" is aroused through the knowledge of cultural norms, it rises into consciousness in considering that certain acts are wrong to pursue and that engaging in them could jeopardize one's honor. Aristotle tells us that when an individual has transgressed a moral code, his fear of social disgrace is exhibited through blushing, as is the case with Coriolanus.[29] Cominius's proclamation, that Martius "must not be the grave of [his] deserving," for "Rome must know/ The value of her own," has the effect of producing the hero's fear of shame and arousing his blush (1.10.20–21). Praise excites negative attention to Coriolanus's martial conduct, provoking pain in his wounds (now the alleged signs of corrupt honor) and causes his face to redden with the guilty knowledge that his deeds had not been accomplished for Rome. The hero's rejoinder, that Cominius's praises are "sauced with lies," further reveals his blush to be a moral reaction aroused through his awareness of transgression (53). Accordingly, the blush reinforces Martius's connection to the Republic and its ideological codes, in turn moderating his pride.

Coriolanus's blush thus calls into question the citizens' earlier statement that he is not "soft-conscienced." Although Martius tells Cominius, "I have done what you have done . . . induced as you have been, that's for my country," his patriotic claim is problematized by his following sentiment: "he that has but effected his good will/ Hath overta'en mine act" (1.10.15–19). As he invokes his commitment to *virtus* and then contradicts his previous assertion, Coriolanus evinces a heightened state of ethical consciousness surrounding his limited mental condition in battle. In accordance with their "will," other men fought more honorably for Rome and demonstrated a higher degree of moral governance over their martial actions. As he reflects upon his misdeeds, Coriolanus's "soul aches" with pain as it gains insight into the material corruption previously imposed upon it (3.1.110). The hero's fear of shame thus begins to generate a conflicting degree of moral awareness surrounding his choleric temperament and proud actions, allowing him to perceive the inherently vicious nature of his passions as well as their debilitating effect upon the will. In this regard, Coriolanus's experience of shame recalls the emphasis early modern divines placed on certain rational passions in their attempts to "limit the consequences of materialist readings of disposition, and in effect repurify a soul that has been . . . saturated with humors, blood, and bile."[30]

Dissatisfied with the potential heresies embedded in humoral theory, medical theologians sought to emphasize man's role as a free agent whose will gave him the capacity to govern his actions. In their writing they also advocated the corresponding view that the rational soul was a free and independent entity and, therefore, could not be overmastered or exonerated in any way by the influence of the body's fluids. As Bishop Edward Reynolds observes:

> The soul depends not on the body in its operations, but educeth them immediately from within itself, as is manifest in the reflection of the soul upon its own nature, being an operation separable there-from, the

> soul being not only *actus informans*, a form informing for the actuating of a body, but *actus subsistence* to, a form subsisting without any dependence upon matter.[31]

As part of their agenda to free the soul from the determinist implications of humoral embodiment, theologians endorsed specific passions of the reasonable soul lodged in the will, including shame, which were instigated through the reflexive operation of conscience and involved in the re-formation of the subject.[32] Timothy Bright observes that the affection which "moveth blushing is shame rising upon the conscience of some known offense," while Reynolds notes how shame, aroused through conscience of evil and guiltiness of mind, "makes us ever reflect upon our weakness." For Reynolds, the blush is the "sign of a mind virtuously disposed in rectifying the quick apprehensiveness of our own defects."[33] By "acquainting themselves with motions," such as shame, "whereunto considered their own determinate qualities have an essential reluctance," Reynolds observes how individuals could undo the "damage and reproach" wrought upon the soul by the body.[34] Generated through the cognitive operation of conscience, Coriolanus's blush motivates a process of moral consciousness and reform through which his soul is released from its determined place in the body. Purged of the choleric fluids that would corrupt and weigh down its immortal nature, the hero's soul is free to act in accordance with his will as a governing principle over the body.

Despite the high level of ethical awareness the hero's blush reveals, blushing also necessarily stands out as an overt revelation of moral transgression. As Lodowick Bryskett explains, "the minde finding what is to be reprehended in us seeketh to hide the fault committed, and to avoid the reproach thereof, by setting that color on our face as a maske to defend us withal."[35] Coriolanus cannot ignore the contradictory quality of his blush as a colorful epidermal border that reveals as much as it conceals. In telling him, "the blood upon your visage dries," Cominius calls attention to the hardened blood on the hero's face as a mechanism of concealment, which appropriates the defensive function of the blush protecting Martius from dishonor (1.11.94). Once the hero moves into the political realm, however, and traditional militaristic tactics of facial concealment are no longer an option, Coriolanus inevitably finds himself overcome by bashfulness, the only self-protective mechanism available. In its earliest classical formulation bashfulness was chiefly understood in relation to the inhibition of unethical action. When placed on the Aristotelian scale of the mean between extremes, modesty could escalate into an excessive shamefastness, in which one's fear of shame could become so overwhelming that it could restrain an individual from engaging in action that would cause disrepute.[36] Spinoza follows Aristotle in his account of bashfulness, which he explains as an extreme "fear that shame will occur, and hence a state of mind restraining man from committing an act which may provoke shame."[37] However, as the concept of bashfulness moved into the wider space of human socialization it progressed from a state of feeling that could curb unethical behavior into a moral emotional condition characterized by extreme behavioral inhibition. As Aristotle observes, men feel shame most "before those who will always be present," as well as before "those not aware of one's misdeeds."[38] In certain social settings the fear of dishonor could progress into an extreme case of modesty, or "moral shyness," in which the subject defensively attempts to hide his *misdeed* or any sign of wrongdoing from view.[39]

As Coriolanus enters the political domain, his behavior perfectly reflects the Aristotelian notion that bashfulness is a moral response rising, as Bright observes, "upon a guiltiness in conceit. This conceit causeth us to hide ourselves, and to withdraw our presence from the society of men, whom we fear do view our faults in beholding us."[40] The senate of patricians is eager to advance Coriolanus to consul. Through such a promotion they seek to "gratify his noble service that/ Hath thus stood for his country," but when Cominius recounts his deeds, "Coriolanus rises and offers to go away" (2.2.38–39). That the hero is experiencing an extreme form of shame—or moral shyness—is made explicit as one senator retorts, "never *shame* to hear/ What you have nobly done" (2.2.65–66; emphasis mine). The ethical thought process intrinsic to shame threatens to reveal itself through the blush. Burdened by the awareness that he did not act nobly on behalf of his country, Coriolanus finds himself overcome by the force of his "shy conscience" and self-defensively withdraws from the room in bashful fear and inhibition.[41] In accordance with political custom Coriolanus must reveal his wounds to the people to prove his worth for consul; yet, he refuses to "put on the gown, stand naked and entreat them/ For my wounds' sake . . . it is a part I shall blush in acting" (2.2.135–137). The central purpose surrounding the display of the wounds is to hold them open to the "rectorship of judgment" so that the people can scrutinize the hero's "marks of merit, wounds received for's country" (2.3.201, 159). In taking the moral implications of the hero's martial conduct into consideration, I suggest here that Coriolanus is ashamed of his wounds because he fears they will be negatively judged as signs of heroic self-interest and vainglory. The wounds are problematic because they provide an opening into Coriolanus's body, materially externalizing the vicious pride and choleric fluids breeding about his heart. As David Hillman reminds us, in the humoral system "virtue and sin were imagined as literally inhabiting one's innards . . . and important subjective and moral truths were understood to lie hidden within the body."[42] In the marketplace, the evaluation of Coriolanus's "marks of merit" would arouse his sense of shame causing him to blush. Lest there is any doubt upon inspection of the wounds the hero's red face would publicize his guilt and confirm his transgression, thus jeopardizing his honor in the Republic and causing the consulship to be taken from him.

On account of his immense pride, Coriolanus is often figured alone as an isolated entity cut off from the social environment. However, by virtue of their paradigmatic openness, Coriolanus's wounds destabilize the sense of sovereignty and enclosure he strives to maintain. Indeed, the very troubling fact that the wounds are open bodily orifices renders Coriolanus subject to a complex affective experience born at the interstices of the self and the sociocultural world he inhabits. Aroused through the knowledge that the wounds signify his transgression of an ideological code, Coriolanus's blush is itself a psychophysiological response in which mind and body interact with the cultural environment in a dynamic action of "continuous reciprocal causation" to produce moral judgments which become reflexively embodied in emotional experience.[43] The red blush of shame thus highlights the skin's role as a porous bodily border where the private self intersects with the world around it. As a permeable somatic marker, Coriolanus's blush parallels the wounds on the surface of his skin, aptly demarcating the distinctive openness of the Galenic self and its capacity for humoral and passional change through its physiological connection to the environment. As he anticipates blushing in the marketplace and suffering an incriminating moment of disgrace, Coriolanus is able to imagine himself as a denigrated object of spectacular appraisal and so begins to consider his body from an abstracted perspective.[44] As he reflects upon his misdeeds and how those deeds come to be problematically revealed through his flesh, the hero's rational soul begins to govern the impulses of his body, as his bashfulness restrains him from engaging in social interaction and corporeal display. In view of the Aristotelian link between bashfulness and humility (bashfulness being the extreme of *the fear of dishonor*; humility, the deficient opposite of its corresponding state, the *desire for honor*) we might note how the "napless vesture of humility" supports the hero's state of modesty when he finally does appear in the marketplace.[45] Functioning as a protective covering, the gown acts as a means of bodily restraint and control that Coriolanus uses to inhibit his wounds from view. As he tells the Plebs, "I have wounds/ To show you which shall be yours in private" (2.3.72–73). Coriolanus has always held an attitude of pride and "noble carelessness" toward the Plebs (2.2.14); however, standing before them in bashful inhibition, the hero's extreme fear of shame and social dishonor reveals that he actually cares about the people's opinion, and so discloses his consideration of their status in the Republic as a party with considerable power.

Believing themselves "mocked," the Plebs misread Coriolanus's performance as a reflection of his pride. The tribunes reaffirm the Plebs' assessment asserting that Coriolanus "wore his humble weeds" with a "proud heart," and convince them to revoke the consulship (2.3.148–49):

Put him to choler straight. He hath been used
Ever to conquer and to have his worth
Of contradiction. Being once chafed, he cannot
Be reined again to temperance. Then he speaks
What's in his heart. (3.3.25–29)

As the tribunes note, Coriolanus gives full reign to his passions. Volumnia has instilled within her son a rigid adherence to his temperament. The hero's ungoverned emotions in battle have been tolerated in view of the safety he provides for the state and excused through the dictates of the Galenic body. However, within the Republic's political sphere Coriolanus's commitment to his choleric nature is no longer tenable, and the tribunes take advantage of his inborn pride and anger as a means of dissolving the consulship before he escalates into a "power tyrannical" (3.3.65). If Coriolanus is to have any chance at regaining the consulship Volumnia must undo her son's essentialist stance by cultivating his sense of temperance. She thus tells him, "I have a heart as little apt as yours,/ But yet a brain that leads my use of anger to better vantage" (3.2.30–31). This is a notion that Coriolanus appears to require more aid comprehending, despite the ethical consciousness and restraint he demonstrated in the last act. Endeavoring to remain constant to his temper, he asks Volumnia, why "would you have me false to my nature?/ Rather say I play the man I am" (3.2.14–16). As Volumnia explains, Coriolanus can dissemble courtesy and humility before the Plebs while preserving his own truth. "Go to them," Volumnia instructs him:

> Thy knee bussing the stones—for in such business
> Action is eloquence, and the eyes of the ignorant
> More learned than the ears—waving thy head,
> Which offer thus, correcting thy stout heart,
> Now humble as the ripest mulberry. (3.2.75–83)

Volumnia's theory of performance is meant to activate her son's sense of temperance through its representation of the rational soul as a governing entity functioning independently of the body. In acting the head is not passively influenced by the physiological operations of the heart; rather, it holds a superior degree of control over the body's fluids, "correcting" the ills of nature.

Learning to dissemble inevitably involves Coriolanus in an act of self-division that causes him to reflect upon his nature and past actions. If performing humility conveys an intrinsically virtuous nature—that Coriolanus fought on behalf of Rome and is now willing to yield to the people as evidence—his realization that he is playing a role "false to [his] nature" must shed light into his choleric passions and the selfish quest for honor that they motivated. Accordingly, the moral discrepancy underlying dissimulation—to be conscious of complete truthfulness while expressing a lie—has the potential to turn Coriolanus's prospective acting into a scene of shame. Because consciousness of transgression is a precondition for arousing the hero's fear of shame, falsely dissembling humility before the Plebs "is a part" Coriolanus will "blush in acting." Like a bad actor he will be unable to dissociate his inner thoughts and feelings from the role he is playing. Envisioning himself dissembling, Coriolanus perceives how his "body's action [will] teach [his] mind/ A most inherent baseness" (3.2.122–123). The bow of humility he enacts will excite his sense of shame, and its conflicting network of moral assessments will cause him to blush, revealing his offence. Yet the revelation of the hero's transgression through the blush will also necessarily disclose the strength of the social bond that ties him to the Republic and its values, adding a level of meaning and sincerity to his kinesic discourse in a way that blurs the line between art and nature. In feigning humility Coriolanus will blush and actually become humble as the ripest mulberry.

As he contemplates how the moral thought process that shame engenders will give way to the blush, the hero imagines his immortal soul, now in the process of being purified, ascending from the corruption of the body. Referring to his body from an abstracted perspective, Coriolanus notes how "this mould of Martius" dematerializes, scatters like "dust . . . against the wind" (3.1.110, 3.2.105). Since, as Bishop Reynolds notes, the soul's dependence upon the "body is not so necessary and immutable, but that it may admit of variation and be vindicated from the impression of the body," Coriolanus further envisages himself undergoing a radical complexional alteration.[46] His "throat of war" once "choired like a drum" turns into "a pipe/ Small as the eunuch or the virgin's voice," and a "harlot's spirit" possesses him (3.2.113–116). As an expression of his "variable complexion," Coriolanus aligns himself with a conglomerate of feminized identities either lost to shame or protected by it (2.1.208). Forced into the social and political landscape and growing increasingly conscious of his transgression, Coriolanus's fear of shame motivates a process of spiritual and moral reform that is itself augmented by humoral shifts in the body. Threatened by the revelation of his offense through the blush, the hero's bashful behavior mimics the retraction of his heart as melancholic humors invade it causing a "withdrawing of spirit and blood, as in fear . . . so that the necessity being more urgent, the blood and spirit break forth again more vehemently and fill the face [with] that shamefast color."[47] Generated through cognitive assessments and perceptual images, the influx of humors coursing through and displayed on Coriolanus's flesh prompts a productive interaction between body, mind, and world, which join together to produce the recurring feeling state of shyness. The perpetual shift in melancholic humors becomes habitual to the extent that it activates a drastic transformation in Coriolanus's mental, emotional, and physical condition. Any irreligious endorsement of the humors as fixed and ungovernable, and the soul as a predetermined, predestined entity, is thus rendered untenable through Coriolanus's experience of shame, which fuses a theological view of the passions as spiritually reformative phenomena with a belief in the Galenic self as a variable organism continuously refashioned as the body comes into contact with the external world.

Though overcome with wounded pride and anger upon his expulsion from Rome, Coriolanus appears to evince a much "gentler spirit" when he arrives in Antium. Submitting to the enemy, the hero proclaims that "he cannot live but to thy shame/ Unless it be to do thee service" (4.5.100–101). Coriolanus's fear of dishonor inclines him to offer his services as a means of nullifying Aufidius's hateful opinion of him. Detecting the hero's modesty, Aufidius wraps his arms "about that body where against/ My grained ash an hundred times hath broke,/ And scarred the *moon* with splinters" (4.5.106–109; emphasis mine). The change in Coriolanus's disposition is more obvious in the Volscian camp, where, as Aufidius claims, Coriolanus:

> Watered his new plants with dews of flattery,
> Seducing so my friends; and to this end
> He bowed his nature, never known before
> But to be rough, unswayable, and free. (5.5.22–26)

In his treatise on bashfulness, Plutarch suggests that shy men who are sensitive to social opinion often feel compelled to "yield to the demands and requests of every man" and conduct themselves in a "servile and flattering manner" so as to gain another's esteem.[48] Offering to serve Aufidius and having "fluttered [his] Volscians," Coriolanus demonstrates a striking resemblance to Plutarch's bashful man—a likeness made all the more apparent when he complies with his mother's request to spare his attack on Rome (5.6.115).

Although Menenius doubts Volumnia's ability to make Coriolanus relent, believing there is "no more mercy in him than there is milk in a male tiger," Volumnia is able to successfully restrain her son's vengeance by employing the logic of shame both gesturally and rhetorically (5.4.28). She first commands the women to kneel: "down ladies; let us shame him with our knees./ To his surname 'Coriolanus's longs more pride/ Than pity to our prayers" (5.3.169–171). Virgilia's curtsy and Volumnia's bow are meant to be expressions of the "fear or respect" inherent in shyness that "caused one to draw back before another," and by doing so, "augmented the portion of that person."[49] However, such gestures of social humility appear to work reciprocally provoking Coriolanus's fear of shame and consciousness of wrongdoing, weakening rather than countenancing his pride. No longer "invulnerable" to "shame," the bashful hero withdraws before his family, declaring, "sink, my knee, i'th'earth" (5.3.73, 50). Coriolanus's state of moral shyness is further evinced when, upon rising, he praises the virgin Valeria by calling attention to the ethical nature of her blush as a mark of feminine virtue. In a recurring image of identification, Coriolanus compliments "the *moon* of Rome, chaste as the icicle/ That's candied by the frost from purest snow/ And hangs on Diane's temple" (5.3.65–67; emphasis mine).[50] The full meaning behind the hero's lyric address is subsequently realized through the counter-arousal of his own blush.

As Carlin Barton observes, "a bond or shared identity awakened one's sense of shame."[51] Volumnia thus endeavors to evoke the relational value of shame through her rhetorical appeals to patriotic and familial bonds:

> Alas, how can we for our country pray,
> Whereto we are bound, together with thy victory,
> Whereto we are bound?
> [. . .]
> thou shalt no sooner
> March to assault thy country than to tread—
> Trust to't—on thy mother's womb
> That brought thee to this world. (5.3.109–111, 123–126)

Coriolanus cannot now ignore the feeling of being bound to the Roman people—nor to Volumnia, whose love and approval he seeks. Sensing his mother's disapproval, the hero's fear of dishonor before her arouses his bashfulness as he "turns away" his face to shield his blush from view (5.3.169). Seizing her opportunity, Volumnia penetrates her son's shy conscience explaining to him that if he destroys Rome he would lose not only familial but public honor. As she declares, "if thou conquer Rome, the benefit/ Which thou shalt thereby reap is such a name/ Whose repetition will be dogged with curses" (5.3.143–145). Deeply aware of his place within a web of human relations and of the ethical implications of his attack, the hero's fear of shame and dispositional bashfulness function to "curb the will" of this most noble Roman, restraining his vengeful course of action and prompting him to yield to his mother's request (3.1.41). Virtuously motivated by the social meaning inhering in Roman notions of honor and *virtus*, Coriolanus displays a moderated sense of pride and restores his fame in the Republic by mercifully and selflessly sacrificing himself for the preservation of Rome.

NOTES

1. William Shakespeare, *Coriolanus*, ed. R. B. Parker (Oxford: Oxford University Press, 1998). All quotes from the play are from this edition.
2. Gail Kern Paster, *The Body Embarrassed: Drama and the Disciplines of Shame in Early Modern England* (Ithaca, New York: Cornell University Press, 1993), 100.
3. Ewan Fernie, *Shame in Shakespeare* (New York: Routledge, 2002), 213, 215.
4. I borrow Douglas Trevor's term "hard-line Galenism" to refer to accounts of the humors and humoral complexion as fixed and irreducible: "Sadness in the *Faerie Queene*," in *Reading the Early Modern Passions: Essays in the Cultural History of Emotion*, ed. Gail Kern Paster, Katherine Rowe, and Mary Floyd-Wilson (Philadelphia: University of Pennsylvania Press, 2004): 240–252; esp. 241.

5. Robert Burton, *The Anatomy of Melancholy* (London, 1652; London: Chatto and Windus, 1883), 1.5.307.
6. Gail Kern Paster, "The Humor of It," in *A Companion to Shakespeare's Works*. Vol 3, ed. Richard Dutton and Jean Howard (Oxford: Blackwell, 2005): 47–66; esp. 48.
7. Lawrence Babb, *The Elizabethan Malady* (East Lansing: Michigan State Press, 1951), 13.
8. Babb, 16.
9. Gail Kern Paster, *Humoring the Body: Emotions and the Shakespearean Stage* (Chicago: University of Chicago Press, 2004), 9.
10. Thomas Wright, *The Passions of the Minde in Generall* (London, 1601), 112.
11. John Sutton, *Philosophy and Memory Traces* (Cambridge: Cambridge University Press, 1998), 40.
12. Wright, 7–8.
13. On the relationship between rational governance and the Aristotelian mean see Joshua Scodel, *Excess and the Mean in Early Modern English Literature* (Princeton, New Jersey: Princeton University Press, 2002). Scodel's study neglects an important strand of Aristotelian thinking about emotions, which I develop in this essay, as involved in the production of rational consciousness and discipline and, hence, not entirely antithethical to reason.
14. Michael Schoenfeldt, *Bodies and Selves in Early Modern England* (Cambridge: Cambridge University Press, 1999), 10.
15. Schoenfeldt, 3.
16. Sutton, 40.
17. Plutarch, *The Lives of the Noble Grecians and Romans*, trans. Thomas North (London: 1579), 237.
18. Eve R. Sanders, "The Body of the Actor in *Coriolanus*," *Shakespeare Quarterly* 57 (2006): 387–412; esp. 405, 411.
19. Coppelia Kahn, *Roman Shakespeare: Warriors, Wounds, and Women* (New York: Routledge, 1997), 148.
20. On Coriolanus's essentialism see Jonathan Dollimore, *Radical Tragedy: Religion, Ideology and Power in the Drama of Shakespeare and his Contemporaries* (Durham, North Carolina: Duke University Press, 2005), 218.
21. Janet Adelman, *Suffocating Mothers: Fantasies of Maternal Origins in Shakespeare's Plays* (New York: Routledge, 1992), 147.
22. Geoffrey Miles, *Shakespeare and the Constant Romans* (Oxford: Clarendon Press, 1996), 154.
23. Trevor, 244.
24. Paster, "Humor," 50, 52.
25. Trevor, 241–242. On the irreligious implications of Galenic theory see also James D. Redwine, "Beyond Psychology: The Moral Basis of Jonson's Theory of Humour Characterization," *English Literary History* 28 (1961): 316–334.
26. Schoenfeldt, 9.
27. Adelman, 149.
28. Aristotle, *The Nicomachean Ethics*, ed. David Ross (Oxford: Oxford University Press, 1998), 2.7.1107b17.
29. Aristotle, *Nicomachean Ethics*, 4.9.1128b12, 2.8.1108a31. For additional work on Aristotelian shame see David Konstan, *The Emotions of the Ancient Greeks: Studies in Aristotle and Classical Literature* (Toronto: University of Toronto Press, 2006).
30. Trevor, 243.
31. Edward Reynolds, *A Treatise of the Passions and Faculties of the Soule of Man* (London, 1640), 404.

32. I develop Trevor's claim in "Sadness in the *Faerie Queene*" concerning religious emotions as associated with conscience and involved in theological attempts to undermine notions of bodily determinism. On the passions of the reasonable soul as distinct from those of the sensitive soul see Wright, 56; Reynolds, 36.
33. Timothy Bright, *A Treatise of Melancholy* (London, 1586), 170–171; Reynolds, 295, on blushing, 275, 311. On the early modern link between shame and conscience see Christopher Tilmouth, "Shakespeare's Open Consciences," in *The Renaissance Conscience*, ed. Harold Braun and Edward Vallence (Oxford: Blackwell, 2011): 67–81; Valerie Allen, "Waxing Red: Shame and the Body, Shame and the Soul," in *The Representation of Women's Emotions in Medieval and Early Modern Culture*, ed. Lisa Perfetti (Gainsville: University Press of Florida, 2005): 191–210; Brian Cummings, "Animal Passions and Human Sciences: Shame, Blushing and Nakedness in Early Modern Europe and the New World," in *At the Borders of the Human: Beasts, Bodies and Natural Philosophy in the Early Modern Period*, ed. Erica Fudge, Ruth Gilbert, and Susan Wiseman (Hampshire, UK: Palgrave, 2002): 26–50.
34. Reynolds, 76.
35. Lodowick Bryskett, *A Discourse of Civill Life*, ed. Thomas Wright (Los Angeles: San Fernando Valley State College, 1970), 176. For related observations see Werner Gundersheimer, "Renaissance Concepts of Shame and Pocaterra's *Dialoghi Della Vergogna*," *Renaissance Quarterly* 47 (1994): 34–56; esp. 47.
36. Aristotle, *Nicomachean Ethics*, 2.8.1108a31, 4.9.1128b12.
37. Benedict de Spinoza, *The Ethics*, ed. Seymour Feldman (Indianapolis, Indiana: Hackett, 1992), 3.31.147.
38. Aristotle, *The Art of Rhetoric*, Loeb Classical Library, ed. J. H. Freese (New York: Putnam, 1926), 2.6.1384a.
39. On the concept of moral shyness see Nathan Rotenstreich, "On Shame," *The Review of Metaphysics* 19 (1965): 55–86; esp. 57.
40. Bright, 170.
41. The divine William Charke notes, "a shy conscience cannot abide her own offences," in *A Treatise Against the Defense of Censure* (Cambridge, 1586), 255. The phrase continues to reflect the moral nature of shame and shyness, as well as their association with conscience throughout early modernity.
42. David Hillman, *Shakespeare's Entrails: Belief, Scepticism and the Interior of the Body* (New York: Palgrave, 2007), 18. Compare with Lear's wish to "anatomize Regan" so that he can "see what breeds about her heart" (*King Lear* 3.6.66–67). For related claims about the bodily interior see Cynthia Marshall, "Wound-Man: *Coriolanus*, Gender, and the Theatrical Construction of Interiority," in *Emerging Subjects: Feminist Readings of Early Modern Culture*, ed. Valerie Traub, Lindsay Kaplan, and Dympna Callaghan (Cambridge: Cambridge University Press, 1996): 93–118; esp. 100.
43. Andy Clark, *Being There: Putting Brain, Body, and World Together Again* (Cambridge, Massachusetts: MIT Press, 1997), 163. For additional work on cognitive neuroscientific approaches to the way in which brain, body, and world interact to produce emotions with specific bodily feelings involved in moral reasoning see Antonio Damasio, *Descartes' Error: Emotion, Reason, and the Human Brain* (New York: Putnam, 1994). The blush, according to Damasio, would be a kind of "somatic marker" that assists in cognitive deliberation and rational decision making (173). On the cognitive neuroscience of emotion, specifically in relation to Aristotle's social constructionist view of emotions as an amalgam of cognitive and biologically based processes, see Konstantinos Kafetsios and Eric LaRock, "Cognition and Emotion:

Aristotelian Affinities with Contemporary Emotion Research," *Theory Psychology* 15 (2005): 639–656. These authors suggest that psychological affections are concurrent with physiological reactions, which in turn work as a means of "amplification," strengthening moral response patterns (651–653). On the blush of shame as a physiological reaction that combines feeling and evaluation and enhances moral reasoning see Steven M. Parish, "The Sacred Mind: Newer Cultural Representations of Mental Life and the Production of Moral Consciousness," *Ethos* 19 (1991): 313–351.

44. The abstracted image Coriolanus creates of himself blushing corresponds to Damasio's intriguing idea of the body as a theatre for the emotions. One need not have to experience the effects of an emotion in the body proper, simply the "fainter image of an emotional body state" can have moral import" (155).
45. In this sense, bashfulness and humility are developed in the *Nicomachean Ethics* as parallel extremes of shame that moderate excessive pride or love of honor. On the Aristotelian idea that passions can moderate opposing passions see *Nicomachean Ethics*, 2.9.1109a30–1109b18. This view of emotions was especially amenable to theologians who endorsed the rationalizing force of certain passions in their struggle against the deterministic implications of humoralism. See Reynolds's description of this view (52). For additional work on the Aristotelian idea of the passions as governors, see Unhae Langis, "*Coriolanus*: Inordinate Passions and Powers in Personal and Political Governance," *Comparative Drama* 44 (2010): 1–27. On the moderating relationship between pride, shamefastness, and humility see Robert Lanier Reid, "Spenser and Shakespeare: Polarized Approaches to Psychology," in *Shakespeare and Spenser: Attractive Opposites*, ed. J.B. Lethbridge (Manchester, UK: Manchester University Press, 2008): 79–120.
46. Reynolds, 8.
47. Bright, 168.
48. Plutarch, "Of Naughty Bashfulness," in *The Morals*, trans. Philemon Holland (London: 1603), 164, 166.
49. Carlin A. Barton, *Roman Honor: The Fire in the Bones* (Berkeley: University of California Press, 2001), 202–203. Barton provides an account of the way shame and shyness operated in the classical world, offering a useful discussion on the socially cohesive nature of the blush (228).
50. The lines may be compared with: "whose blush doth thaw the consecrated snow/ That lies on Dian's lap" (*Timon* 4.3.388–389).
51. Barton, 207.

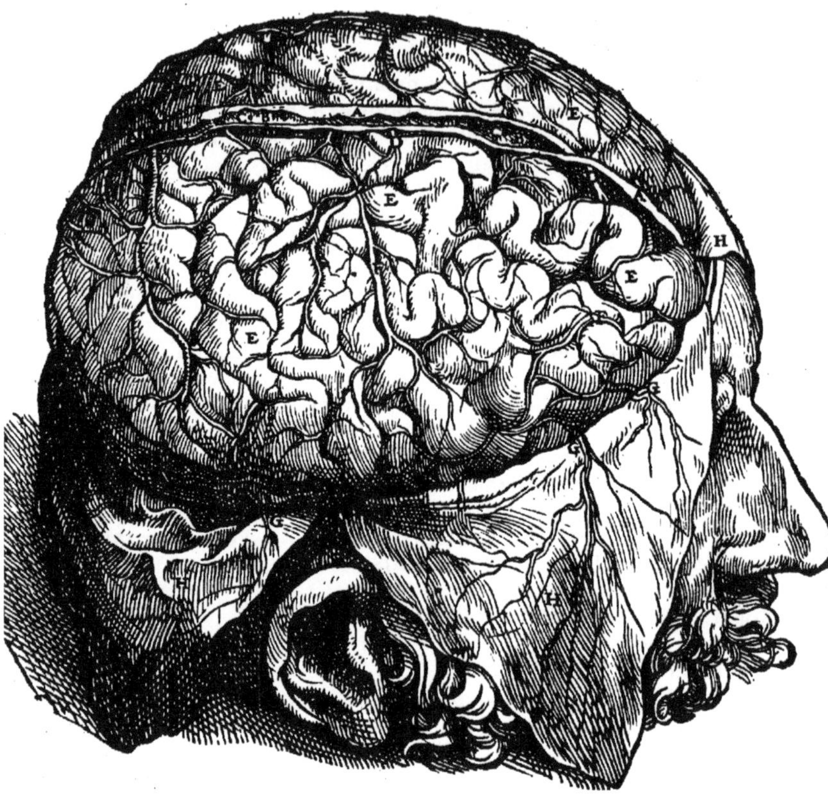

Figure 5 "Brain Plate," from Vesalius, *De humani corporis fabrica* (1543), 606 (Liber VII). Image A027122, History of Medicine Collection, Courtesy U.S. National Library of Medicine.

[...Fourth Link...]

The Play of Time in Cognition

Katherine Rowe

These thoughts are positioned at a moment of shifting scale. Essays offering local readings, that bridge contemporary theories of distributed cognition and early modern literary studies cede place to metacritical reflection on the challenges and opportunities the former offers to cultural historians now. For readers who drop into the volume at this moment, that shift of scale is likely to spark speculation and reflection well in advance of the metacritical essays themselves. Why do Renaissance texts answer so well to the demands of distributed cognitive theory? What are the most compelling ways in which an extended mind theorist might frame the past in this comparative analysis? Even more interesting and perhaps more difficult to grapple with is how does the passage of time play in cognitive systems?

Early modern theatre texts and the performance practices that shaped them were scripted by ways of knowing prior to our own—most crucially, to the post-Cartesian disengagement of matter from cognition. For this reason, these works offer us a salutary form of estrangement from ways of thinking about knowing we have inherited from seventeenth-century English philosophers and the Enlightenment writers who followed them. As John Sutton reminds us, post-Cartesian dualism reconfigured the body as "a static, solid container" for the mind, "only rarely breached, in principle autonomous from culture and environment"—an arena of executive control and volition "tampered with only by diseases and experts."[1] Since the seventeenth century, the imperative to investigate that rarely breached territory has generated diverse arenas of expertise: from early forensic anatomy to Reformation soteriology to Freudian psychoanalysis and the many disciplines of self-knowledge that have followed.

For the legacy thinkers that we are, early modern texts provide a salutary corrective to dematerialized and individualized models of cognition. In this volume, they are the occasion for a complexly historicized invitation: to use an emerging model of cognition to revisit a historically distant one in order to glean new evidence from familiar archival fields. Think backward from the Enlightenment focus on sight as a privileged mode of perception, for example, to an early modern preoccupation with the more distributed sense of touch (the governing or "common sense" of early anatomies): a host of new meanings leap out of the ubiquitous haptic metaphors of Shakespeare's plays.

From recognizing skin as an edge and container for body, we may revisit skin as a material extension of thought, as interface, as transactional system.

To study cognitive activities longitudinally in this way is to come to see ways of knowing as historically conditioned. Arnold Davidson and Lorraine Daston have labeled this mode of cultural history "historical epistemology": the study of changing categories of knowledge; of the cultural preconditions that make thinking in a certain way possible; of the material conditions and practices in which such categories emerge, change, disappear. The critical imperative, as Daston describes it, is to "understand the history of the categories that structure our thought, pattern our arguments and proofs, and certify our standards for explanation."[2] Critics pursuing this comparatist historical enterprise typically find ourselves grappling, from moment to moment, with the inherent anachronism of our practice; should our local claims be for historical analogy, continuity, distance, or disjunction? Methodologically, the enterprise requires all of these claims, since the ways that humans conceive of internal processes develop continuously but unevenly and can endure over long spans of time. As Dirk Geeraerts and Stefan Grondelaers have argued, the dominant categories for emotion in the West involve a kind of conceptual entail from Galenic theory, "part of the lexical legacy of the theory of the four humors."[3] Humoral theory in turn embedded prior metaphorical understandings of bodily experience (such as "anger as fire") drawn from earlier classical systems.[4]

A model of distributed cognition poses a special challenge and opportunity for cultural historians, inviting us to think more systematically about the role of the past in ways of knowing. The opportunity is twofold: first, to understand how our own investigative and analytic practices are constrained by the structural entails of dualist thought, such as we find self-reflectively mapped in Evelyn Tribble's evolving work on the Globe Theatre as a "smart structure," part of a distributed cognitive system; and second, to understand more clearly how the passage of time plays in such systems, as Ed Hutchins's influential claim that cognition is "a fundamentally cultural process" implies.[5] How do *processes* of thinking carry historical entails? The social and material scaffolding by which humans enskill, offload, or underload builds temporal and cultural lag into cognitive processes. All "smart structures"—the Globe's distinctive architecture, the roll that stores a player's part, prevailing gestural vocabularies for suiting action to word, genre protocols—are necessarily polychronic ones. Thus agency, in cognitive systems, must be understood as distributed diachronically as well as synchronically.

This recognition complicates any explanation of what it means that the scaffolding of distributed cognition which *once* worked on an audience's imaginary forces in the early seventeenth century *still* works. In his 1948 film of *Hamlet*, Laurence Olivier directs "To be or not to be" in ways that strikingly recall the engravings of Thomas Willis. Discovering Hamlet's mental pathology for the audience via montage (Figure 6), he makes a cinematic case for the superior forensic power of film to penetrate Hamlet's mystery:

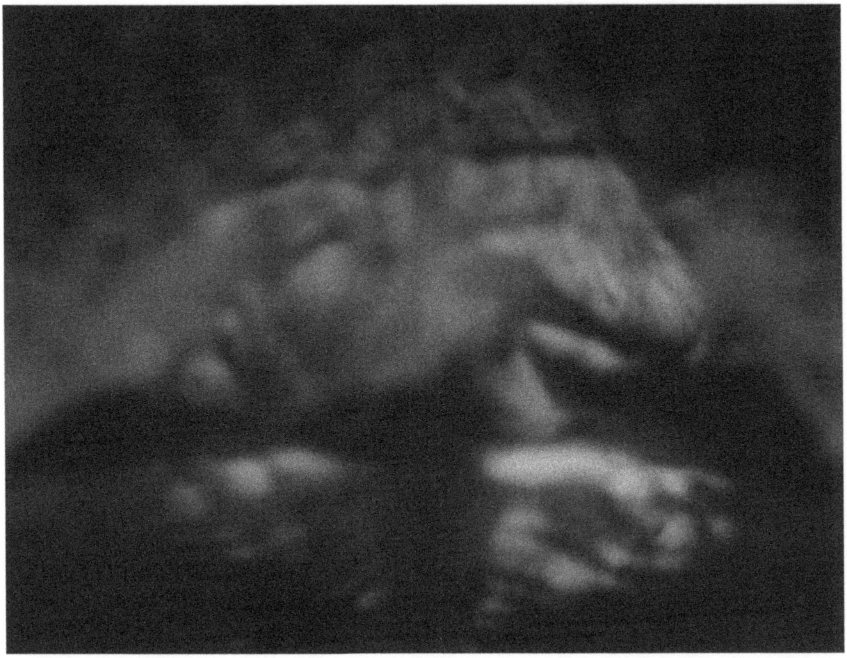

Figure 6 "Hamlet's brain in forensic montage," *Hamlet* (1948; dir. Laurence Olivier).

In this montage, and in an opening voice-over ("This is the tragedy of a man who. Could not. Make up. His Mind"), Olivier reprises a topos (investigation of the brain) and forensic imperative (diagnose cause by looking inside) that seems uncannily to be called forth by the analytic drives Lianne Habinek traces here in the play's revenge-plotted brain. Yet for Olivier and his audience, that topos and those genre protocols carry fundamentally different understandings of mental impression, sufferance, and agency. (Including the implied notion that a mind *can* be "made up" wilfully, requiring Olivier to invent his new tagline.) Every apparent congruency like this one raises follow-on questions of noncongruency: why do some cognitive scaffolding systems (forensic *in* and *out* in theatrical media for example), appear especially portable over time, while others do not? That is the kind of question that will only be answerable by means of interdisciplinary analyses bridging the domains of study pertaining to the imbrication of cultural practices, bodies, and minds.

[...]

NOTES

1. John Sutton, *Philosophy and Memory Traces: Descartes to Connectionism* (Cambridge: Cambridge University Press, 1998), 41.

2. Lorraine Daston, "Historical Epistemology," in *Questions of Evidence. Proof, Practice, and Persuasion across the Disciplines*, ed. James Chandler, Arnold I. Davidson, and Harry D. Harootunian (Chicago: The University of Chicago Press, 1994): 282–289, esp. 282.
3. Dirk Geeraerts and Stefan Grondelaers, "Looking Back at Anger: Cultural Traditions and Metaphorical Patterns," in *Language and the Cognitive Construal of the World*, ed. John R. Taylor and Robert E. MacLaury (Berlin and New York: Mouton de Gruyter, 1995): 153–180; esp. 170.
4. Geeraerts and Grondelaers, 172.
5. Ed Hutchins, *Cognition in the Wild* (Cambridge, Massachusetts: MIT Press, 1995), 374.

10 Altered States

Hamlet and Early Modern Head Trauma

Lianne Habinek

"About, my brains!"

> At all times when a man is about to commit any thing in custody to his *Memory*, first let him study to drown all unnecessary thoughts in oblivion, that he may perfectly intend the things he is to learn.
>
> —John Willis, *Mnemonica; Or, The Art of Memory*[1]

John Willis's 1618 memory treatise advises its student to take a course of action strikingly similar to Hamlet's infamous pledge to wipe from the table of his memory "all trivial fond records,/ All saws of books, all forms, all pressures past/ That youth and observation copied there" (1.5.99–101).[2] The underlying idea—which is not without its discontents—is that a blank slate allows one all the better to entrust to the custodian Memory the important things requiring remembrance. Willis continues:

> *Oblivion* being such a principle of *Memory*, as *Privation* is of *Generation*; and a ready remembrance most commonly proceedeth from right understanding the thing in hand; therefore a man must prepare himself diligently, and so untie the force of his imagination, that he may as it were engrave and imprint occurent things in his Memory.[3]

For Willis, the effacement preparatory to imprinting involves unlinking imagination from memory, thereby negating its potentially distorting sensory and intellective influence so that memory can focus on the "thing in hand," allowing present occurrences to "engrave and imprint" themselves upon the mind. But how should one go about effecting such an erasure, and where, physically, should that erasure occur? We are of course now aware that the seat of memory is the brain, and our current understanding of the functions and specific localizations of memory's component parts derive from decades of study involving lesions. In such studies, scientists make inferences about the state of a healthy brain by observing perturbed behaviors in subjects with specific (known or intentionally created) types of brain damage.[4] Work of this sort has been invaluable for mapping cognitive function onto brain location, albeit tentatively and with numerous caveats.[5] This method of studying the brain has its roots, I argue below, in the broader history of neuroscience. Moreover, in the absence of modern imaging technology, these roots are steeped in violence and trauma to the brain. Drowning the active, sensing mind in oblivion or wiping clean a mental tablet is not an innocent action.

Criticism of *Hamlet*'s fascination with the human mind and body has often focused on Yorick's skull as a memento mori upon which Hamlet meditates as he attempts to answer his seminal question, "To be or not to be?"[6] Here, I am more interested in what is (or rather, was) *inside* the skull (indeed, all the skulls in the play). I want to suggest that in *Hamlet* the violence of a willed oblivion inaugurates investigations that seek to reconstruct perceived damage to memory, to self, to brain, and to state in order to understand the nature of such damage and how a healthy or normally functioning system would operate. In short, *Hamlet* is distinctive in its concern with the brain and with what transpires should that brain be damaged in some way. And, as we shall see, the particular investigations carried out in the play have close metaphorical connections to those carried out by contemporary surgeons and anatomists.[7] A curious thing about *Hamlet* is that every instance of the word *brain* has to do with the organ being ill or shaken or emptied or mistaken. In other words, the brain is constantly seen to be fallible in the play, an entity forever malfunctioning and which no amount of thought or action seems to be able to set quite right. Horatio, for example, fears that the Ghost may put "toys of desperation" into the young prince's brain (1.4.75–76); Hamlet offers to wipe his brain so that the Ghost's commandment will live there (1.5.98–103); Hamlet rails against his own brain for making him so pigeon-livered as to lack gall to stand against his uncle (2.2.522); Claudius worries that Hamlet's brain puts him "from fashion of himself" (3.1.173–174); Gertrude, after Hamlet has seen the Ghost which she has not, fears that it was a hallucination, a "coinage" of his brain (3.4.135). The human brain in *Hamlet*, then, is always apt to misfire—and it is crucial that the brain in question is invariably Hamlet's.[8]

The wounds involved in *Hamlet* are evident from the play's inception: the opening question, "Who's there?" presumes an absence, a hole in the proper functioning of a system (i.e., if the machine of perception and cogitation worked fluidly, it would be obvious who was there). What, then, can be utilized to work through this lesion, to solve the "crime" of the damage, to restore order? Two central mysteries in the play—the murder of Old Hamlet, and young Hamlet's madness—constitute both real and metaphorical wounds which, in turn, require Hamlet in the first place, and Claudius in the second, to attempt to study and understand the extent of the damage and then to reason backward in order (ostensibly) to restore justice or health.[9] One investigation, then, involves the work Hamlet must do to test the truth of the Ghost's testimony: in order to "catch the conscience of the king," Hamlet stages *The Mousetrap* as a kind of theatrical laboratory which "comes near the circumstances" of Old Hamlet's death, allowing him to recreate in a controlled space and manner the initial scene of trauma (i.e., Claudius's poisoning of Old Hamlet; 2.2.540; 3.2.72). My reading here focuses instead on the initial traumatic event that triggers the other investigation in the play, namely that of Claudius (and by extension the rest of the Danish court) into Hamlet's mind. Whether real or feigned, the prince's madness occurs after his visitation from the Ghost and after he pledges to wipe his memory so as to inscribe the Ghost's commandment in the "book and volume" of his brain. I want to focus on this pledge, for it is, I argue, a self-inflicted wound, a violent act that damages Hamlet and renders his memory and brain a locus of forensic investigation for Claudius.

To focus the interpretation of this play on the brain further allows us to see its indebtedness to an increasingly important area of study in early modern medicine. Physicians and anatomists working in the sixteenth and seventeenth centuries evince a kind of anxiety about the location or the locatability of the soul that appears remarkably similar to our modern attempts to localize, for example, semantic memory in the brain's hippocampus[10] or certain types of emotion in the limbic system.[11] As an increasing number of early modern thinkers have argued, the location of the soul was the human brain—and the brain therefore was not the domain of science and medicine alone; rather, it was the site not only where body and soul connected but also where various disciplines (divided today by nearly insurmountable boundaries) productively converged. Indeed, the work of early modern physicians describes an interesting link between experimentation and an understanding of how the mind and body connected via the brain, a link that *Hamlet* exploits. To probe the play in this manner demands, though, a brief contextualization, by considering the kind of investigation that contemporary surgeons and anatomists were conducting when faced with damaged brains. This historical context will provide a framework both for my reading of *Hamlet* as well as for thinking about the emergent practice of neuroforensic investigation.

"PRIVATE ERROUR"

There are quacks abroad, warns physician Franciscus Arceus in a 1588 treatise. Something is rotten in the state of medicine when "unlearned and unskilfull Chirurgions" practice upon patients: "obscure and unperfect experience, is joyned with most perfect knowledge," laments Arceus, such that these hapless pseudo-doctors "know nothing els then that which they have learned of their Maisters of the same sorte . . . everie one of them doth follow his owne sense or private errour, having no author at all, although never so unlearned to lead them thereunto."[12] Arceus's key complaint here has to do with the imperfect transmission of knowledge and experience, either via incompetent teachers or through the faults of the students themselves. Indeed, it is the impossibility of distinguishing student from master that causes, in part, such a collapse of authority, for if "every one of them doth follow his owne sense," then where is the original, where the appropriate method of medicine? Transmission errors destroy the proper growth of intelligence; for such errors to be "private" is for them to be like a festering wound, or disease, of which the observable symptoms are not merely "covetousnesse," "want of judgement," or "arrogancie of minde," but also "lacke of knowledge."[13] Such serious "errour" erodes the characters of the would-be practitioners—keeping them too content in their work, too dull—but, naturally, it also has serious consequences for the patients unlucky enough to fall into their faltering hands.

It is not by chance that Arceus's admonishment appears at the start of a text on how to treat head wounds. Indeed, the purpose of *A most excellent and Compendious Method of curing woundes in the head* is to staunch this damaging ignorance, to lay bare a method that will curb the tide of "errour" which leads to madness (and worse) in mishandled patients. For the prognosis for head wounds in the seventeenth century does not seem to have been particularly rosy. "Grievous wounds are either *Periculosa*, or *Malefica*, or *Lethalia*, or *Difficilia*," notes the celebrated physician Alexander Read in one of his lectures (1650) at the Barber-Surgeons Hall: "*Periculosa*, or dangerous wounds, are such as sometimes are cured, although for the most part they bring death; such [are] superficiall wounds of the braine."[14] Read cautions that damage to the brain is naturally very dangerous because of the organ's preeminence in the body; "the wounds of the braine . . . are deadly, if they penetrate to the basis; because it is a principall part, the well-spring of the animall facultie, in continuall motion, and because inflammation, and a sharpe fever insueth."[15]

Arceus would have been acutely aware of these potential complications. His panacea for this "errour" is method, the careful accounting of a system, the laying out of a standard procedure that will not be corrupted by ineptitude or deviated from by those who think they know better. He rebukes such authorless practitioners and students for their failure to consider the wounds they encounter with anything like order or analysis. The unmethodical physician, according to Arceus, examines only the surface, not caring to dig deeper. Such physicians

> doe neglect to serch out throughlie whether any thing be hurt or perished in the *Rete Mirable*, or any of the other panicles or compactions of the braine, for the partes of the lower or innermost bone . . . oftentimes happeneth to be cut in sunder, shivered, dashed, & broken in peeces, and moved out of their places, and that fault is found more oftener in the inner Table, then in the upper. Which thinges first most grievous panges and griefes, and after death it self doth ensue.[16]

The "most grievous panges and griefes," as well as death, follow not merely from injuries to the brain, but from the investigative failures of quack surgeons, the "private errour(s)" turned horribly public.

But what could be done to investigate the "inner Table" of the brain, particularly in the late 1500s, when Arceus was writing? After all, we have only recently worked out how to peer into the interior of the living, functioning brain. The error of the injured brain would seem to be quite private, indeed—so private, in fact, as to be unintelligible even to the person who suffered the wound. The diagnosing physician could only hope to understand the extent of the injury in one of two interrelated ways. Neither of these would grant him access to the original error or fault, but they would at least allow him to investigate obliquely. First, he could attempt to recreate the circumstances of the wound in the hopes of discovering the nature and extent of the damage. Second, he could observe the patient's behaviors—which here would include suffering "grievous panges and griefes" and then death—and then make guesses about which parts of the brain had been affected. Such a method represented an early means of scientific systematization: the physician could diagnose the troubled mind by stripping away complications to get at the essential truth of the wound. Anatomist Helkiah Crooke describes a similar methodology, whereby one must reason backward from effect to cause in explaining how we may use anatomy to discover the nature of God: "all the knowledge of God that can be had," he writes, "must be derived . . . not from any cause or matter preceding, but from the effects and thinges subsequent."[17] Thus when Crooke argues that the brain, rather than the heart, is the principal member of the body and so responsible for motion and sensation, he does so by contrasting what happens when something adversely affects each organ. Our own experience, he writes, suggests that

> when the ventricles of the Braine, are either compressed, or filled and stuffed up, as in the Apoplexy, Epilepsie, and drowsie *Caros*, then all the faculties are respited and cease from their functions; but when the heart is offended, the life indeede is endangered, but neither motion nor sense intercepted.[18]

Here we can glimpse a rudimentary version of the technique widely employed by modern cognitive scientists in order to understand the brain's functions—namely, the lesion study. In an actual lesion study, cognitive scientists usually recruit patients who have—through disease, head trauma, or necessary operations—lesions or damage to a specific section of their brains. A well-known example of a long-term lesion study with only one subject was the work done with Henry Gustav Molaison (patient HM), who in 1953 had surgery to treat severe epilepsy; doctors removed a significant part of his medial temporal lobe, which houses the amygdala and hippocampus, areas of the brain associated with memory, emotions, and navigation. Knowing the location and extent of the damage to Molaison's brain, those working with him could design cognitive tests to determine the extent of the damage to his memory (Molaison had significantly impaired short-term memory)—and, in so doing, could theorize the roles of those areas in normal cognition.[19] Examining the cognitive abilities of lesion patients could thus potentially allow the investigator to map these abilities onto particular areas of the brain in a normal patient.

While the imaging techniques available today are radically new, the conceptual and methodological principles underlying such studies are much older even than Crooke. Their roots may be traced to the work of the foundational figure in western anatomy, Galen, whose ideas concerning the brain derived from his work with injured gladiators.[20] Galen was perhaps the first in western practice to link the idea of the mind or soul and its attributes to the human brain; what is more, he was probably the first western physician to study the effects of trauma to the brain and nervous system and thus to begin to draw the lines between specific functions and specific sections of the brain.[21] As surgeon to the gladiators of Pergamum, Galen witnessed at first hand alterations in sensation, movement, and thought that might follow when an unlucky competitor sustained various sorts of head trauma.[22] The deeper and more serious the wound to the brain, the more severely affected the gladiator's behavior. Through his observations, Galen even managed to discover correlations between the location of the brain damage and the sort of behavioral change that followed: a laceration to one of the brain's lateral ventricles altered motion and sensation, while damage to the third ventricle could garble reason.[23] In one infamous public demonstration, Galen countered his cardiocentric detractors by slicing the laryngeal nerve of a squealing pig, arguing as the pig fell silent that the brain and nervous system, because of their direct influence from the soul, had far more control over the body than did the heart.[24]

Note that it would have been extremely difficult for Galen to theorize links between brain and mind *without* resorting to damaging the former to consider the effects upon the latter. In the absence of functional magnetic resonance imaging (fMRI), the early physician had to rely on what he could study using his own senses—and naturally there was no way to observe directly the soul acting on the brain and the brain acting on the body. In Galen's work, then, are the roots of a very modern conception of cognitive experimentation, namely the idea that the brain must be studied by looking at cases of things that have gone wrong with it. So, too, the treating of the head wound required method. Arceus argues that one should first discern the cause of the wound: was the instrument sharp or dull, did the patient fall or get hit, what were the patient's mental and physical states at the time of the injury, how strong was the person who inflicted the damage, and so on. Next one must examine the skin surrounding the injury; if necessary, the skin should be broken in order to expose the extent of the damage to the skull, and parts of the skull should then be removed to give the brain room to breathe. All of this must be undertaken *before* treatment (which involved applying poultices to the wound), so that the very nature of the injury could be observed and recorded. That is, one must question circumstances; one must make the damage worse before one can make it better; one must remove various obstacles to the access of the original wound; one must treat only when one understands the nature of the wound.

Arceus holds himself to this standard of medical investigation. He discusses a case wherein his patient was involved in the construction of a tower adjacent to a church in Valnerde; a stone a cubit in length and width struck the man on the head and fractured further his sagittal sulcus. His people bore him home, where "he suffered the passion of the braine three whole daies together, not onely speechles, but also without any mouving, & was moved & turned of the standers by no otherwise then if he had bene dead."[25] Blood issued from his nose and eyes, and, what is more, his neck and head were black and swollen. The physicians worked round the clock and observed that:

> the third day he spake, yet unperfectlie, and as men of a troubled minde are wonte to doe, his eyes were open and staring, in the manner of them that have the falling sicknesse, or that lie in a traunce, but he saw nothing at all, nor yet began to see untill the xx. day, but after that hee mended every day in his sight. And after the second moneth, he could see verie well as when he was in health. . . . And he rose presently out of his bed, allthough he could not then well goe, and so by the helpe of God more then by the helpe of man, he was healed. And he escaped his eies looking a squint. One, one way, and another, another way, the which also (as the cure did proceede) at the fourth moneth were restored and looked right, he liveth as yet, and hath married a wife. For he was then a yong man.[26]

Following such a careful method of treatment, Arceus argues, will lead to the recovery, health, and eventual normalcy of your patient. Their young man was treated according to such dictates, and not only did he not lose his sight, he also escaped his "eies looking a squint" and was able to marry—a prognosis considerably more positive than that of a patient not treated according to this method.

Arceus and Read both argue that the aim in such medical investigation is a cure. The brain should be allowed to be purged as a last resort—it is much healthier and safer (or so the thinking went, though today we would be skeptical: how safe can it be to feed a ten-year-old small beer while applying a poultice of juniper and thyme to his open brain?) to staunch the wound, to clean it, to probe it, but to try to heal it. Arceus's method is a rudimentary version of the kind of empiricism that would, later in the century, be posited by Bacon and others as a new means of accessing essential natures. England's first dedicated neuroanatomist, Thomas Willis, for example, would note that the difference between the brain of a normal person and that of a "fool" would be apparent in anatomical dissection, as the cerebellum of the former (Figure 7) would be large and healthy, while that of the latter (Figure 8) would be thin and small.[27] Although it is now taken for granted that brains are the receptacles of minds and are responsible for thought, memory, and the like, this was by no means a given in premodern times. Galen was among the first to make the connection during his work with wounded gladiators, and it was upon this simple yet seminal idea that the method of Arceus, modern neuroscience, and—as I am arguing—*Hamlet* is based.

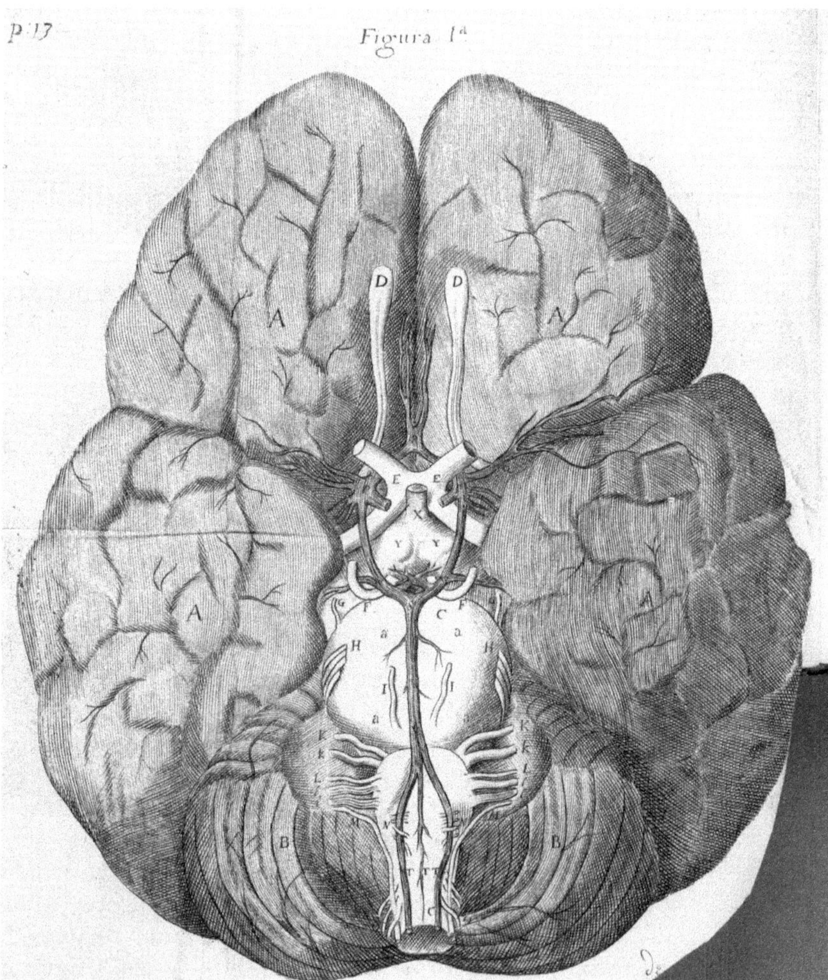

Figure 7 Thomas Willis, *Cerebri anatome* (London, 1664), figure 1a (a healthy cerebellum), fold-out plate, 12. Image reproduced by permission of The Huntington Library, San Marino, California.

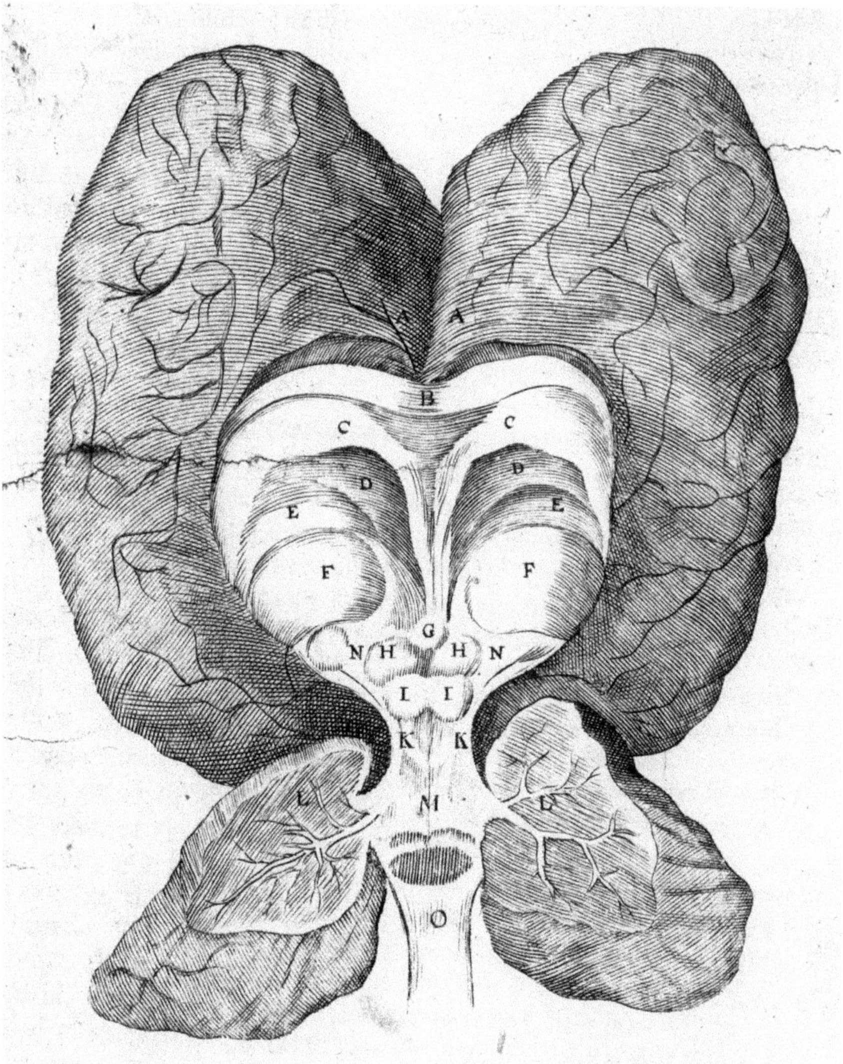

Figure 8 Thomas Willis, *Five Treatises* (London, 1681), figure 4 (the cerebellum of a "foolish youth"), opp. 70. Image reproduced by permission of The Huntington Library, San Marino, California.

"MY TABLES"

The method proposed for treating head trauma in the early modern period requires a careful reconstruction of the events surrounding the initial incident of trauma. For a reading of *Hamlet* in this context, the pertinent first question is, then, where and how this first traumatic incident occurs. I focus my reading here on one such incident, namely Hamlet's pledge to wipe his memory so that he may remember solely the Ghost's "commandment." This becomes the traumatic event it will be Claudius's self-appointed responsibility to investigate:

Ham. Remember thee?
Yea, from the table of my memory
I'll wipe away all trivial fond records
All saws of books, all forms, all pressures past
That youth and observation copied there,
And thy commandment all alone shall live
Within the book and volume of my brain,
Unmix'd with baser matter.
[. . .] Now to my word.
It is "Adieu, adieu, remember me." (1.5.97–111)

Key to this scene is the linking of brain to book and the depiction of memory via an extended metaphor of writing. But what is the specific capacity and function of this mercurial object? From the perspective of textual history, there are a few striking things about how this passage treats the brain. Intriguingly, the brain itself does not appear in the First Quarto (1603):

Yes thou poore Ghost; from the tables
Of my memorie, ile wipe away all sawes of Bookes,
All triviall fond conceites
That ever youth, or else observance noted,
And thy remembrance, all alone shall sit.[28]

In fact, Q1 simplifies both the method and the object of Hamlet's declaration: youth and observance merely "noted" the "sawes" and "triviall fond conceits," so they may be the more readily wiped away, and it is not the Ghost's "commandment" but simply his "remembrance" which will "sit" in Hamlet's "tables." The action here is decidedly uncomplex—erase the tables and reinscribe a new remembrance. This "remembrance" would allow Hamlet more latitude of recollection; he might have space in his freshly wiped tables to accommodate both the Ghost's description of Old Hamlet's murder and the Ghost's call to revenge.[29] Yet any sense of the location of this wiping away has itself been effaced from the text.

The elaboration of both the Second Quarto (1605) and the Folio (1623) versions of this speech to include reference to the brain (as well as to "sinnows" and "sinnews," respectively, which refer to nerves) underscores a related transformation effected by these later texts: a shift in the locus of memory from the heart to the brain. George Williams points out that one source for Hamlet's metaphor can be found in the Geneva Bible's Book of Proverbs. Proverbs 7:1–3 urge the wise person to "kepe my wordes, and hide my commandements. . . . Kepe my commandements, & thou shalt live. . . . Binde them upon thy fingers, and write them upon the table of thine heart."[30] Here the memory is kept in the heart, as words and commandments written upon its tables. "The word of the Father, clearly, must supersede all 'trivial fond records,/ All saws of books . . .' etc.," remarks Williams; "that commandment—or the specific part of it Hamlet chooses to record—is 'Remember me'" (141).[31] A similar anatomical metaphor occurs in Heywood's 1607 *A Woman Killed with Kindness*, in which Wendoll, who is charged by his friend Frankford to seduce Frankford's wife Anne, claims that her refusal of him is "recorded/ Within the red-leav'd table of my heart."[32] Ann and John Thompson note the similarity of this passage to Hamlet's promise: "the permanence of the printed or written, as opposed to the transitory nature of spoken language, is relevant and the 'coveredness' of heart and brain, like the 'coveredness' of books, is seen as helpful."[33] Such a preference for "coveredness" recalls, too, the injunction of Proverbs not just to "kepe my wordes" but to "hide my commandements"—writing on both heart and brain is thus secret, "visible" only to the shared memory of he who commands and he who is commanded. In Hamlet's case, as we shall see, this "coveredness" will slip dangerously into solipsism as the prince fails to transcribe his father's commandment properly onto a memory that has been wiped clean of everything.[34]

What is key about the two examples I have cited above is that both indicate the heart, rather than the brain, as the table book upon which memories will be transcribed. This point, coupled with Q1's lack of anatomical referent, highlights the uniqueness of the later versions of Hamlet's declaration: the writing will take place in or on the brain. The focus of the later versions of *Hamlet* on the brain in this passage is part of the play's broader fascination with the brain. Here, the brain is not yet malfunctioning: this passage marks the moment of trauma from which Hamlet's madness will proceed. The question will become, however, whether the trauma can rightly be located in the wiping of the memory or in the memory's inscription with the Ghost's commandment. Both options involve writing—erasure on the one hand and reinscription on the other. The metaphor of writing is key to thinking about memory's relationship to the brain and its repercussions should that memory be "wiped." The wax metaphor of Aristotelian tradition was still useful for describing how memory worked in the early modern period; it was particularly attractive to those seeking to link anatomy and humoral psychology.[35] Excessive heat or coldness would damage the brain—and subsequently the memory, as the organ would be rendered more or less impressionable—in much the same way as wax would be affected by similar temperature gradients. Laurentius's discourse about vision explains:

> Experience also giveth us to understand, that if the braine have his temperature altered: as for example, if it be too hot, as it falleth out in such as are franticke: or over cold, as it falleth out in melancholick

> men; it corrupteth presently the imaginative facultie, troubleth the judgement, weakeneth the memorie. . . . [The soul] requireth a good temperature.[36]

A "good temperature" will enable the soul to operate as it is meant to do; coldness in particular will weaken the memory, although we are not here given the precise mechanism. Melancholy itself affected memory, as Laurentius points out, whether the cold creates the melancholy or the melancholy the cold. Rhodri Lewis, citing Timothy Bright's humoral description of melancholy, argues that were melancholy affecting Hamlet, he "would not have been able to think in an orderly fashion; his imagination . . . would have exercised a dominant influence over his words, thoughts, and behavior."[37] Thus John Willis's suggestion that a man wishing to remember something "untie the force of his imagination" from his memory would have been most prudent for the melancholic.

The wax metaphor was attractive for early modern theorists because, as Stephen Greenblatt puts it, it conveyed "subtle variations, gradations, and effacements of impressions, some cut deeply, virtually permanently, into the medium of memory, others only lightly, mostly slowly losing their original distinctness of outline and eventually fading altogether."[38] John Willis more elaborately considers the mechanism by which temperature affects the memory:

> Imbecillity of Natural Memory proceedeth from too much heat, coldness; moisture, or dryness of the brain. Overmuch heat wasteth and consumeth animal spirits; too much coldness obstructeth motion of the Spirits lodged in the Cells of the Brain; superfluous dryness causeth such callocity, that the species cannot be imprinted: Redundant moisture doth erase and obliterate forms of things, as soon as they are imprinted.[39]

The animal spirits—manufactured in the brain from the vital spirits in the heart, and key tools of the soul working in the human body[40]—may have their motion obstructed by changes in heat and moisture and, as such, are unable to carry out what John Sutton terms "wriggle-work," a sort of engraving motion of the animal spirits upon the human brain.[41]

Yet there seems to be something more than wax on the prince of Denmark's brain. As Peter Stallybrass, Roger Chartier, John Mowery, and Heather Wolfe famously argue, wax tables have their limits both as a metaphor and as a stage device: they would not be capable of retaining permanently the Ghost's commandment, nor would they be a feasible prop for an actor to lug around onstage. A book whose pages could be written upon in ink is problematic in that it could not easily be erased or wiped clean. Thus some technology was necessary as a halfway point between these two devices: an instrument which could bear both permanent and temporary traces, preserving the one while allowing for the near-infinite removal and reinscription of the other. In a note to the "tables" speech in the Third Variorum *Hamlet*, James Boswell the younger (son of Dr. Johnson's biographer) remarks that "I am in possession of three of these *table-books*: one printed in 1604, the date of the first edition of Hamlet: 'Writing Tables, with a Kalendar for xxiiii Yeeres, &c.' The Tables made by Robert Triplet. London. Imprinted for the Companie of Stationers, 1604."[42] Though they do not mention Boswell's note, Stallybrass and his coauthors propose that early modern table books underpin the actual metaphor at stake here. These table books generally consisted of printed almanacs bound with pages that had been specially treated so as to allow for inscription and wiping (e.g., with a damp cloth or fingertip).[43] Such tables were easily portable and allowed the carrier to make temporary notes that could later be transferred into, for example, the more permanent record of a commonplace book. Stallybrass and colleagues explain how this particular technology is meant to operate in Hamlet's case:

> When Hamlet uses writing tables as a model of memory, they also suggest forgetfulness: Hamlet wants to forget all that "youth and obseruation" have "coppied" in his mind. His mind's tables must first be wiped clean of all this earlier copying so that a single command can be permanently inscribed there—the Ghost's "Remember me." It is surely because there could be no less suitable technology than erasable tables for a permanent remembrance that Hamlet metamorphoses the "Tables. . .of Memory" into the quite different "Booke and Volume" of his brain, which he imagines as a place of indelible writing.[44]

The key here is that the tables offer Hamlet a space that can be wiped clean, while the "book and volume" of his brain can contain the "indelible writing" of the Ghost's injunction. Thus Hamlet's memory involves a dual system, a marriage of wax and stylus on the one hand and paper and pen on the other.[45]

Some difficulty occurs, however, when we consider that these two writing instruments must coexist within the narrow region of Hamlet's brain. Table book and volume would also need to share access to sensory and intellective information, so that temporary notes from the one might be more permanently transferred into the other. Wiping the contents of the table book to free up space in the "book and volume" suggests that damage to the table book results in parallel damage to the "book and volume." In externalizing the act of memory through the metaphor of writing (even if that writing takes place in the interior space of the brain) the table book is, to borrow a Derridean term, a supplement to memory: a helpful but potentially dangerous tool in that it offers the fantasy, but not the actual means, of permanence. Problematically, it seems that in the act of transcribing the Ghost's "commandment" into his memory, Hamlet ensures he will forget it. As Douglas Brooks argues, "Within seconds of hearing the father's last words . . . the forgetful son suffers a memory lapse the moment he attempts to take dictation"; having sworn his memory to the Ghost, "Hamlet recalls the central analogical tradition for describing the workings of memory in terms of the very exteriorized supplements that memory has come to rely on."[46] A number of modern studies corroborate what Plato demonstrated with the warning fable of the *pharmakon* and what Hamlet proves: as soon as one ceases to actively hold information in one's memory, that information will be lost. Whether one writes down the information, presses a button, speaks, or otherwise communicates the information beyond one's own memory, the neural circuits responsible for maintaining the information in working memory will cease to fire.[47]

As commentators have noted, in writing his "word," Hamlet facilitates his forgetting of it.[48] The additional irony lies in his forgetting precisely the word *remember*, for it is in writing down that the act of forgetting is initiated. Moreover, what is further forgotten in the act of noting down the need to remember is the Ghost's primary charge:

Ghost. Pity me not, but lend thy serious hearing
To what I shall unfold.
Ham. Speak, I am bound to hear.
Ghost. So art thou to revenge when thou shalt hear. (1.5.5–7)

The Ghost would have Hamlet seek revenge, but what the prince uses to overwrite his memory is the injunction *remember me*.[49] The act of remembering, particularly when one has suffered damage to the organ of memory, proves difficult—and this difficulty is compounded when the act of remembering is itself what causes the damage. For what else is there to remember when the Ghost's commandment is the sole living entity in Hamlet's brain? I point this problem out to highlight what I see as the violence inherent in the act of wiping away memory and inscribing a particular commandment. Garrett Sullivan comes close to articulating the effects of such violent forgetting:

> the Ghost's words initiate in Hamlet a fantasy of the annihilation and reformation of the self through forgetting and the subsequent inscription of a memory trace. . . . Interestingly, the forgetting that is to precede Hamlet's inscription would destroy the contents of his memory. Hamlet understands the contents of his memoria and the Ghost's demand that Hamlet remember him as mutually exclusive.[50]

If forgetting is an act of "annihilation," then so, too, must remembering be. William Proctor Williams, for example, notes at least two performances of *Hamlet* in which the actors whip out daggers to assist in the inscription of the Ghost's commandment.[51] The use of daggers here points, too, to what

Jonathan Goldberg has described as the violence of the early modern act of writing; knives were often used in the period to scratch out writing on paper.[52] As Juliet Fleming explains:

> writing, even in ink, can be "erased" ("scraped out") from paper (for that matter, any writing that was not finally erasable would not be writing at all). . . . But such erasures are neither readily effected nor infinitely repeatable: they are possible, but not practical. . . . [T]o cross or "expunge" ("prick out") writing is to mark it, more or less memorably, for erasure: it is, in effect, to *write it again*. Clearly, then, paper does not preserve writing that has been effaced from it.[53]

Even erasing writing, then, is a sort of writing again, or writing over. Inscription of any sort, whether writing or erasing, seems to constitute a kind of violence. One cannot write, remember, or forget without causing damage—to the physical medium of writing or to the psychic medium of the self.

To literalize the play's metaphor by thinking of such inscription as taking place on the brain recalls the physician's inquiry into a case of head trauma. Memory was thought to be housed in the fourth or hindmost ventricle of the brain.[54] Crooke observes that

> it is a perpetuall truth in the body of a man, that by how much the cavity is greater, by so much it is the baser. The fourth ventricle is of all the rest the least and the narrowest, and containeth the Animall spirit sincere defœcated and exquisitely purged: the others do onely prepare the spirit, and therefore the hindmost ventricle is the most noble.[55]

This fourth ventricle, smallest and housing the "exquisitely purged" animal spirit, deploys this spirit in service of memory itself. How else can we be certain that the fourth ventricle is the highest and best? Crooke explains: "*Galen* teacheth that the upper ventricles are the basest . . . by the example of a young man of *Smyrna* a Citty of *Ionia*, who being wounded into one of the upper ventricles yet escaped with life . . . A wound in the hindmost ventricle doth most of all offend the creature."[56] It is through this sort of backward reasoning, the logic and violence of the lesion study, that we are able to comprehend what we do about the properly functioning brain. Hamlet's pledge to remember the Ghost's commandment constitutes an act of self-damage which it will be the task of Claudius to probe. The effect of Hamlet's lesion will be madness—real or feigned, we know not which—that will seem, as Claudius will observe, to be "something in his soul/ O'er which his melancholy sits on brood" (3.1.166–167). Resolving this something, as we know, incurs further violence—a purgation.

At the end of Act 1, Scene 5, the ventricle containing Hamlet's memory has been wiped clean. Upon its surface the injunction to remember is inscribed—but the referent has been lost. The prince's mind and memory become an echo chamber of remembrance without an entity that can properly be said to be remembered. Thus Hamlet delays, and it is not until he encounters Yorick's skull that he is reminded strikingly of the mortality of all things. Throughout the play, but particularly in the graveyard, Hamlet encounters numerous forms of death, to which his living brain responds with wit[57]—but he has forgotten the very "canoniz'd bones, hearsed in death" of his father, and "the sepulchre" which seemed to have "op'd his ponderous and marble jaws" to cast forth not Old Hamlet's body but merely the ghostly semblance of that body (1.4.47–50). In remembering Yorick, Hamlet remembers that he has forgotten the king.

Since we find ourselves in the graveyard, let us turn to what it contains: the crypt. In his essay "Fors," a preface to Nicholas Abraham and Maria Torok's *The Wolf Man's Magic Word*, Jacques Derrida explores the architecture of the crypt, a constituent part of the self created through the mourning process and essential for it. The crypt, he writes, "is not a natural place" but a place "*comprehended* within another but rigorously separate from it, isolated from general space by partitions, an enclosure, an enclave. So as to purloin *the thing* from the rest." Inside itself, the crypt constructs a safe, "sealed, and thus internal to itself, a secret interior" which is "the condition, and the stratagem, of the cryptic enclave's ability to isolate, to protect, to shelter from any penetration, anything which can filter in from outside." Yet "the crypt can only constitute its secret by means of its division, its fracture. 'I' can only *save* an inner safe by putting it inside 'myself.'"[58] Derrida's description returns us to the epigraph with which I began this essay, for it underscores the internal division evoked by Willis's description of the effacement necessary to memory. Following John Willis's rules, a man could force himself to remember by sending to oblivion the rest of his memory and carving out a space untied from other sensory or imaginative inputs. In this sense, the fourth ventricle becomes itself a crypt—in order that, as Derrida puts it, I may "pretend to keep the dead alive, intact, *safe* (*save*) *inside me*, but it is only in order to refuse, in a necessarily equivocal way, to love the dead as a living part of me, dead *save in me*."[59] Hamlet likewise loves his dead father only as a thing living inside himself—he has, after all, promised that the Ghost's commandment "all alone shall live" inside his brain. Such a promise folds dead into living and renders the prince's skull more crypt than cranium.

It is not, then, simply a "something" in Hamlet's soul that troubles him so and that incites the madness which Claudius studies and of which, eventually, he will purge the prince. The "something" sitting over Hamlet's soul is the violence of loss and of memory, the uncoupling of sensory and intellective input from recollection, which causes him to keep his dead father's "word" alive inside himself. Old Hamlet is dead *save in* Hamlet; he lives insofar as his son lives with the memory of death. Such an attempt will disease his wit (3.2.313), blunt his purpose (3.4.111), and eventually allow his dying brain to recognize its mortal wounds. A normal person, facing death, would be able to utter only, "I am dying." Hamlet, carrying all along the violent inscription of the dead upon the living matter of his brain, will be able to speak the unspeakable, "I am dead," for he has been so all along (5.2.338).

NOTES

1. John Willis, *Mnemonica; Or, The Art of Memory, Drained out of the pure Fountains of Art & Nature . . .* (London, 1661), C6v–C7r. This text is a full translation of his 1618 *Mnemonica; sive Ars Reminiscendi: e puris artis naturaeque fontibus hauta. . .*, which appeared in partial English translation in 1621. See Frances Yates, *The Art of Memory* (London: Routledge and Kegan Paul, 1966), 336 n.40.
2. William Shakespeare, *Hamlet*, ed. Harold Jenkins (Surrey: Thomas Nelson & Sons Ltd., 1997). All references will be to this edition, unless noted in the text.
3. Willis, C7r.
4. "Known" lesions are generally those detected by fMRI in human subjects who have experienced trauma or necessary corrective surgery, whereas "intentionally created" lesions are performed upon animal subjects. Foucaud Du Boisgueheneuc, Richard Levy, Emmanuelle Volle, Magali Seassau, Hughes Duffau, Serge Kinkingnehun, Yves Samson, Sandy Zhang, and Bruno Dubois explain the utility of combining such studies to better localize function: "In humans, functional imaging (PET scan and fMRI) cannot show whether one given region of a network is critical (i.e., the cognitive process cannot be implemented without this region) or accessory (i.e., the cognitive process can be partially or fully functional without it). Therefore, lesion studies in humans are particularly useful to complement experimental studies in animals and functional imaging studies." See their "Functions of the Left Superior Frontal Gyrus in Humans: A Lesion Study," *Brain* 129 (December 2006): 3315–3328; esp. 3324. For a particularly interesting example, see Alessandra Geraci, Luca Surian, Marco Ferraro, and Anna Cantagallo, "Theory of Mind in Patients with Ventromedial or Dorsolateral Prefrontal Lesions Following Traumatic Brain Injury," *Brain Injury* 24.7–8 (2010): 978–987. They examine "Theory of Mind," which involves "representing and reasoning about mental states" (such as making inferences about another person's emotional state based on his or her facial expression), in patients with damage to particular parts of the prefrontal cortex following traumatic injury (978).
5. See, for example, the controversy surrounding the potential segregation of spatial and nonspatial working memory in the prefrontal cortex, as reviewed by Clayton E. Curtis and Mark D'Esposito, "The Effects of Prefrontal Lesions on Working Memory Performance and Theory," *Cognitive, Affective, & Behavioral Neuroscience* 4.4 (2004): 528–539. See also Chris Rorden and Hans-Otto Karnath, "Using Human Brain Lesions to Infer Function: A Relic from a Past Era in the fMRI Age?" *Neuroscience* 5 (2004): 813–819. Rorden and Karnath question the utility of the lesion study given the wide range and increasing precision of fMRI.
6. See, for example: Harry Morris, "'Hamlet' as a 'Memento Mori' Poem," *PMLA* 85.5 (1970): 1035–1040; James V. Holleran, "Maimed Funeral Rites in *Hamlet*," *English Literary History* 19 (1989): 65–93.
7. This essay is a part of a larger project regarding what I term "neuroforensic investigation" in *Hamlet*. In brief, I consider the sick, malfunctioning brain

as a metaphor for the two *sites of investigation* in the play: Hamlet's mind and the court of Denmark. The skull is, both literally and metaphorically, a crime scene: in one case, a leprous distillate is poured into Old Hamlet's ears, and, in another case, Hamlet himself feigns madness (which involves the soul within the brain). Two crimes have been committed, two brains disturbed—so two investigators must examine the sick brain empirically. The first three Acts of *Hamlet* involve the *act* of investigation; by contrast, Acts 4 and 5 demonstrate the solution of the crime and the subsequent healing of the wound (in the form of purgation). See Lianne Habinek, "Altered States: *Hamlet* and Early Modern Neuroforensics," in *"Such Wondrous Science": Brain and Metaphor in Early Modern English Literature*, Diss. Columbia University, 2009. (Ann Arbor: University of Michigan, 2011): 111–166.

8. A notable exception is the Player Queen's ominous exhortation that sleep rock the Player King's brain; see 3.2.222. Then, too, upon encountering his mad sister, Laertes wishes that heat would dry up his brains; see 4.5.154.
9. For a boiled-down explanation of the idea that damage to a system can reveal its inner workings, see Amy Cook, "Staging Nothing: *Hamlet* and Cognitive Science," *SubStance* 35 (2006): 83–99; esp. 90–91. Cook's schematic account of V. S. Ramachandran's work on anosognosia (a condition in which a person who has suffered a given problem, like paralysis or blindness, as a result of brain damage is unable to acknowledge that the problem exists) describes as well the "lesion" study of humans.
10. See Joseph R. Manns, Ramona O. Hopkins, and Larry R. Squire, "Semantic Memory and the Human Hippocampus," *Neuron* 38.1 (2003): 127–133; Larry R. Squire and Stuart M. Zola, "Episodic Memory, Semantic Memory, and Amnesia," *Hippocampus* 8.3 (1998): 205–211; and F. Vargha-Khadem, D. G. Gadian, K. E. Watkins, A. Connelly, W. Van Paesschen, and M. Mishkin, "Differential Effects of Early Hippocampal Pathology on Episodic and Semantic Memory," *Science* 277.5324 (1997): 376–380.
11. See James W. Papez, "A Proposed Mechanism of Emotion," *Journal of Neuropsychiatry and Clinical Neuroscience* 7.1 (1995): 103–112; and Paul D. MacLean, "Some Psychiatric Implications of Physiological Studies on Frontotemporal Portion of Limbic System (Visceral Brain)," *Electroencephalography and Clinical Neurophysiology* 4.4 (1952): 407–418.
12. Franciscus Arceus, *A most excellent and Compendious Method of curing woundes in the head, and in other partes of the body, with other precepts of the same Arte, practiced and written by that famous man Franciscus Arceus, Doctor in Phisicke & Chirurgery: and translated into English by John Read, Chirurgion* (London, 1588), Ciir–v.
13. Arceus, Ciir–v.
14. Alexander Read, *A Treatise of the First Part of Chirurgery [. . .]: The part which teacheth the re-unition of the parts of the body disjoyned. Containing The Methodicall Doctrine of Wounds: Delivered in Lectures in the Barber-Chirurgeons-Hall, upon Tuesdaies, appointed for these Exercises, and the keeping of their Courts* (London, 1650), B2r–v.
15. Read, B3r.
16. Arceus, Ciir.
17. Helkiah Crooke, *Mikrokosmographia* (London, 1615), C1v.
18. Crooke, E2v–E3r.
19. See Alan J. Parkin, "H.M.: The Medial Temporal Lobes and Memory," in *Classic Cases in Neuropsychology*, ed. Chris Code, Claus-W. Wallesch, Yves Joanette, and André Roch Lecours (East Sussex: Psychology Press, 1996): 337–348.

20. See John Scarborough, "Galen and The Gladiators," *Episteme* 5 (1971): 98–111.
21. See Edward S. Hodgson,"Long-Range Perspectives on Neurobiology and Behavior," *American Zoologist* 30.3 (1990): 403–505; esp. 416: "Several centuries before Galen, [the Indian physician] Sushruta . . . had discovered that damage to one branch of the recurrent laryngeal nerve of a human resulted in a hoarse voice *and* a loss of taste. . . . The Hindus, who would not dissect pigs (as Galen had done), derived their information from war casualties."
22. See Matthey Tomey, Ricardo Komotar, and J. Mocco, "Herophilus, Erasistratus, Aretaetus, and Galen: Ancient Roots of the Bell-Magendie Law," *Neurosurgical Focus* 23 (2007): E12–E14; and P. M. Dunn, "Galen (AD 129–200) of Pergamun: Anatomist and Experimental Physiologist," *Archives of Disease in Childhood: Fetal and Neonatal* E 88 (2003): F441–F442.
23. The human brain has four ventricles which store cerebrospinal fluid; in the Galenic model, these ventricles housed refined, not-quite-physical spirits, which served the will of the soul. See, for example, Charles Singer, *Vesalius on the Human Brain,* ed. E. Ashworth Underwood (London: Oxford University Press, 1952), 40.
24. See Mirza Baig, Faheem Chishty, Phillip Immesoete, and Chris Karas, "The Eastern Heart and Galen's Ventricle: A Historical Review of the Purpose of the Brain," *Neurosurgical Focus* 23 (2007): E3–E9; Charles G. Gross, "Galen and the Squealing Pig," *The Neuroscientist* 4 (1998): 216–221; and John Scarborough, "Galen Redivivus: An Essay Review," *Journal of the History of Medicine and Allied Sciences* 43 (1988): 313–321.
25. Arceus, Fivr–v.
26. Arceus, Fivr–v.
27. Thomas Willis, *Cerebri anatome* (London, 1664), figure 1a (a healthy cerebellum), fold-out plate, 12; and Thomas Willis, *Five Treatises* (London, 1681), figure 4 (the cerebellum of a "foolish youth"), opp. 70. Images reproduced by permission of The Huntington Library, San Marino, California.
28. William Shakespeare, *The Tragical Historie of Hamlet Prince of Denmarke* (London, 1603), C4v–D1r. Given that one theory regarding the relation of the quarto and folio texts involves memorial reconstruction, it is certainly ironic to consider that the word *brain* might simply have been forgotten by the person transcribing Q1. For opposing viewpoints in the discussion of the textual history of the play, see Paul Werstine, "The Textual Mystery of *Hamlet*," *Shakespeare Quarterly* 39.1 (1988): 1–26; and R. A. Foakes, *Hamlet Versus Lear: Cultural Politics and Shakespeare's Art* (Cambridge: Cambridge University Press, 1993), 90–97. For an intriguing third prong in the debate, and an advocacy of a palimpsestic reading of the print and performance history of the play, see Laurie Johnson, *The Tain of Hamlet* (Newcastle-upon-Tyne: Cambridge Scholars Publishing, 2013), 33–45.
29. Had Hamlet remembered what he intended to remember, we would have a very different play indeed, as evinced by Q1's leanness and emphasis on action rather than contemplation. As Richard Helgerson laments, "Readers sometimes forget what a straightforward play this opening [i.e., of Hamlet remembering what he sets out to remember] would have begot"—"What Hamlet Remembers," *Shakespeare Studies* 10 (1977): 67–97; esp. 70–71.
30. George Walton Williams, "Hamlet's Tables," *Hamlet Studies* 1.2 (October 1979), 141.
31. Interestingly, although Williams made this observation in 1979, he seems to have been alone in doing so: none of the major editions of *Hamlet* (e.g., Hudson,

Dyce, London, Bevington, Pelican, Arden, Oxford, New Cambridge, New Folger, Norton, Riverside, or Guild) note the potential reference.

32. Thomas Heywood, *A Woman Killed with Kindness*, in *A Woman Killed with Kindness and Other Domestic Plays*, ed. Martin Wiggins (Oxford: Oxford University Press, 2008), 6.126–127.
33. Ann Thompson and John O. Thompson, *Shakespeare: Meaning & Metaphor* (Iowa City: University of Iowa Press, 1987), 166.
34. Further examples of the metaphor abound in the footnotes, glosses, and marginalia of the Folger Shakespeare Library's copy of the third variorum Shakespeare, vol. 7 (shelfmark PR2752 1821e c. 2 v. 7). Dante, Montaigne, Dekker, and Beza are among the referents; see William Shakespeare, *The Plays and Poems of William Shakespeare, With the Corrections and Illustrations of Various Commentators: Comprehending A Life of the Poet, and An Enlarged History of the Stage, by The Late Edmond Malone, With a New Glossarial Index*. vol. VII (London: 1821).
35. See, for example, Douwe Draaisma, *Metaphors of Memory: A History of Ideas about The Mind*, trans. Paul Vincent (Cambridge: Cambridge University Press, 2000), 24–27.
36. M. Andreas Laurentius, *A Discourse of the Preservation of the Sight: of Melancholike Diseases; of Rheumes, and of Old Age*, trans. Richard Surphlet (London, 1599), 4.
37. Rhodri Lewis, "Hamlet, Metaphor, and Memory," *Studies in Philology* 109.5 (2012): 609–641; esp. 639.
38. Stephen Greenblatt, *Hamlet in Purgatory* (Princeton, New Jersey: Princeton University Press, 2001), 214.
39. John Willis, L6r.
40. For a clear account of this process, along with a helpful diagram, see Singer, xviii–xx.
41. See John Sutton, *Philosophy and Memory Traces. Descartes to Connectionism* (Cambridge: Cambridge University Press, 1998), 25–49. My reference here is a simplification of Sutton's rich and compelling metaphor.
42. Shakespeare, *Plays*, 249.
43. See, for example, Roger Triplet, *Writing Tables with a Kalender for xxiii. yeeres, with sundry necessarie rules* (London, 1604).
44. Peter Stallybrass, Roger Chartier, John Mowery, and Heather Wolfe, "Hamlet's Tables and the Technologies of Writing in Renaissance England," *Shakespeare Quarterly* 55.4 (2004): 379–419; esp. 415; but see also Lewis's critique of the study, 610–616. A more modern correlate to the table-book certainly haunts the account by Stallybrass et al.—although the authors never mention Freud's "mystic writing-pad," the idea is remarkably similar. See Sigmund Freud, "A Note Upon the 'Mystic-Pad,'" in *Collected Papers*, vol. 5, ed. James Strachey (New York: Basic Books, 1959): 175–180. See, too, P. K. Ayers's assertion that the "trivial fond records" are actually more metaphorical in nature, comprising the "basic structures of scribal education, the compressed summaries of all fields of knowledge which function as the cues needed to retrieve and employ the larger corpus already filed in his mind"—"Reading, Writing, and *Hamlet*," *Shakespeare Quarterly* 44.4 (1993): 423–439; esp. 430.
45. Can the memory willfully be erased in such a matter? Lewis argues that such "mnemonic erasure" is impossible (614, passim), while Sutton suggests that "neurological ethics" in the early modern period allowed philosophers to theorize that we can be taught "to influence or directly act upon our own brains and spirits, to achieve true virtue by the exercise of moral dominion over our own bodily fluids" (16).

46. Douglas A. Brooks, "'Within the Book and Volume of My Brain': *Hamlet* and the Crisis of Memory—Past and Present," *Foreign Literature Studies / Wai Guo Wen Zue Yan Jiu* 28.117 (2006): 24–31; esp. 28.
47. See the following studies for experimental discussion of this fact: John Anderson, Lynne Reder, and Christian Lebiere, "Working Memory: Activation Limitations on Retrieval," *Cognitive Psychology* 30 (1996): 221–256; Shintaro Funahashi, Matthew Chafee, and Patricia Goldman-Rakic, "Prefrontal Neuronal Activity in Rhesus Monkeys Performing a Delayed Anti-saccade Task," *Nature* 365 (1993): 753–756; Henk Haarmannand Marius Usher, "Maintenance of Semantic Information in Capacity-limited Short-term Memory," *Psychonomic Bulletin & Review* 8 (2001): 568–578; and Earl Miller, Cynthia Erickson, and Robert Desimone, "Neural Mechanisms of Visual Working Memory in Prefrontal Cortex of the Macaque," *The Journal of Neuroscience* 16 (1996): 5154–5167.
48. See Ronald Knowles, "*Hamlet* and Counter-Humanism," *Renaissance Quarterly* 52.4 (1999): 1046–1069. Knowles points out, "Hamlet does not realize that [keeping the Ghost's commandment alone in his brain] is impossible. He cannot replace a mind shaped by rhetoric with unalloyed feeling. Rhetoric provided not just knowledge, but *how* knowledge was assimilated and understood: it provided a cognitive structure which enforced the Western censure of emotion" (1064).
49. See John Kerrigan, "Hieronimo, Hamlet and Remembrance," *Essays in Criticism* 31.2 (1981): 105–126. As Kerrigan sagely observes, "Revenge is stifled by remembrance" (119).
50. Garret A. Sullivan, Jr., *Memory and Forgetting in English Renaissance Drama. Shakespeare, Marlowe, Webster* (Cambridge: Cambridge University Press, 2005), 13.
51. "In the Tony Richardson/Nicol Williamson production, the 'My tables' is cut, but 'meet it is I set it down' causes Williamson to 'write' on the stone wall with his dagger. In a similar fashion, Mel Gibson, in the Zeffirelli film, writes enthusiastically on the stones of the parapet with his sword"—William Proctor Williams, "Hamlet's Pockets: Problems with Stage Directions," in *Inside Shakespeare: Essays on the Blackfriars Stage*, ed. Paul Menzer (Selinsgrove, Pennsylvania: Susquehanna University Press, 2006): 192–99; esp. 195.
52. See Jonathan Goldberg, *Writing Matter: From the Hands of the English Renaissance* (Stanford, California: Stanford University Press, 1990).
53. Juliet Fleming, *Graffiti and the Writing Arts of Early Modern England* (Philadelphia: University of Pennsylvania Press, 2001), 75–76.
54. See Stanley Finger, *Origins of Neuroscience: A History of Explorations into Brain Function* (Oxford: Oxford University Press, 2001), 333; T. Manzoni, "The Cerebral Ventricles, the Animal Spirits and the Dawn of Brain Localization of Function," *Archives Italiennes de Biologie* 136.2 (1998): 103–152; esp. 131; and Mary J. Carruthers, *The Book of Memory. A Study of Memory in Medieval Culture* (Cambridge: Cambridge University Press, 1990), 52; 303 n.31.
55. Crooke, Yy6v–Zz1r.
56. Crooke, Yy6v–Zz1r.
57. See Gail Kern Paster, "The Pith and Marrow of our Attribute: Dialogue of Skin and Skull in *Hamlet* and Holbein's *The Ambassadors*," *Textual Practice* 23.2 (2009): 247–265; esp. 261.
58. Jacques Derrida, "*Fors*," trans. Barbara Johnson, *The Georgia Review* 31.1 (Spring 1977): 64–116; esp. 67–68, emphasis in the original.
59. Derrida, 71.

11 *Cogito Ergo Theatrum*

Redistributing Cognition on the Early Modern Stage

Laurie Johnson

Scholars in Shakespeare Studies might be expected to view with suspicion any explanation of culture or society based on models of constraint, having vigorously debated the merits of "subversion" versus "containment" models throughout the historicist turn in this field during the 1980s and 1990s. By *constraint*, we refer to the capacity of cultural, environmental, or social factors to limit or delimit individual actions—the triumph of structure over agency. A model of constraint is in such terms easily aligned with the "containment" model, with which the work of Stephen Greenblatt and the New Historicists has long been associated, extending as it does the approach to culture developed by the anthropologist Clifford Geertz.[1] Typically the work of British Cultural Materialists has been seen as advocating the "subversion" model, albeit one in which subversion can only be identified on the basis of material evidence that it was perceived and acted upon, as such, by those in positions of authority and power.[2] I raise these debates at the beginning of this chapter because the specter of the "subversion" versus "containment" debate is one old mole that I imagine being bound to rise from the cellar as we attend to questions of body and mind in relation to the early modern stage. My primary focus here is an examination of the model of "distributed cognition" initially outlined within Early Modern Studies by Evelyn Tribble in her 2005 essay, "Distributing Cognition in the Globe"[3] and more fully developed by her in 2011 in *Cognition in the Globe: Attention and Memory in Shakespeare's Theatre*.[4] I demonstrate that studies into distributed cognition reduce the risk of recapitulating a model based on constraint, which is not to admit that a version of this approach in Shakespeare Studies would automatically slide away from a containment model towards a recapitulation of the subversion model. I argue that the more recent material cultural histories of early modern practices have already shown the way forward for a cultural materialism that is not anchored solely in concerns for the vicissitudes of power. By drawing on a rich material cultural history to construct her evidence base for the examination of early modern stage practices in terms of distributed cognition, Tribble compels us to reconfigure both text and performance in Shakespeare's theatre in terms of the early modern body-mind.

To begin, we should clarify why the terms *constraint* and *containment* are being described here as having sufficient overlap in meaning that explanations offered on the basis of the one may lead us to imagine a return to the other. *Constraint* is a key defining feature in attempts by modern social psychologists to reconfigure the relative causal importance of social factors on individual behavior. Whereas W.H.R. Rivers had argued almost a century ago for the union of sociology and psychology on the basis that all social activities have mental states as their source,[5] subsequent work on the interplay between the two fields has produced the mantra of the primacy of the social field: as Peggy Thoits explained in 1995, "sociological mechanisms are crucial for explaining key psychological phenomena and . . . structural contexts constrain individuals' behaviors in ways often overlooked by psychologists."[6] Fifteen years earlier, in outlining the method that he called a "poetics of culture"—"New Historicism" would be the more widely accepted name for this method within a few years—Greenblatt presented a quite similar series of statements: "Social actions are themselves always embedded in systems of public signification"; "Language, like other sign systems, is a collective construction"; "Self-fashioning for such figures involves submission to an absolute power or authority situated at least partially outside the self."[7] Greenblatt's model is anthropological, culture being defined *pace* Geertz as "a set of control mechanisms . . . for the governing of behaviour,"[8] in which we thus find a constraint model residing at the heuristic core of his approach. When, in "Invisible Bullets," Greenblatt concludes that "genuine and radical" subversiveness "is at the same time contained by the power it would appear to threaten,"[9] he extends this heuristic focus on the constraint of individual behaviour—operating at the level of structure *or* agency—into an argument about containment in which the controlling structure (in the service of power) is seen as functioning in the mode of a defence mechanism *against* subversive agency.

It would be easy to dismiss the specter of the subversion-containment model as no longer having any residual impact in Shakespeare scholarship. By 1993, amid a raft of critiques of Greenblatt's work, Linda Charnes famously referred to the "stalled subversion-containment model," and the idea seemed already to have had its day.[10] Greenblatt himself appeared to cast off the New Historicist yoke during the next decade, as both *Hamlet in Purgatory* and *Will in the World* saw the founder of New Historicism shifting his focus toward literary biography.[11] Yet in 2002, a collection of scholarship inspired by Greenblatt's appearance at the "Touch of the Real" seminar in 1998 appeared in print.[12] While the essays were hardly reiterations of Greenblatt's methods across the board, nevertheless all were responding after a fashion to his renewed illustration of the New Historicist method in the context of an appraisal of Geertz's influence in "The Touch of the Real," published in *Representations* in 1997.[13] A year earlier, Patricia Parker's

Shakespeare from the Margins: Shakespeare, Culture, Context sought to go beyond the "now largely abandoned" strain of New Historicism that had spawned subversion-containment, yet she aligned her reading of "the fault lines of apparently orthodox utterances" with the Raymond Williams-inspired Cultural Materialists, thereby situating this work within the terms of the earlier debates.[14] In one of the more programmatic moments of this book, Parker observed that to read Shakespeare in terms of social order and power, the reader must also "include within any conception of a political Shakespeare an awareness of the language that both stages this order and subversively dismantles it."[15] Indeed, in his most recent book on Shakespeare, Greenblatt returns from the opening comments to the question of agency and constraint, seemingly having somewhat turned around on the subject:

> Shakespeare as a writer is the embodiment of human freedom. He seems to have been able to fashion language to say anything he imagined, to conjure up any character, to express any emotion, to explore any idea. Though he lived his life as a bound subject of a monarch in a strictly hierarchical society that policed expression in speech and in print, he possessed what Hamlet calls a free soul. . . . Yet if Shakespeare is the epitome of freedom, he is also a figure of limits. These limits are not constraints on Shakespeare's imagination or his literary genius. Doubtless there were such constraints—notwithstanding his aura of divinity, he was, after all, a mortal—but I am among those who are struck rather by the apparently unbounded power and visionary scope of his achievement.[16]

If the subversion-containment model seemed to have stalled by 1993, then the terms through which it framed scholarship of Shakespeare's social or cultural context nevertheless remained a focus of numerous studies throughout the 1990s and into the new millennium. This ongoing concern with structure or agency *qua* subversion-containment within Shakespeare Studies is worth a watchful eye as we examine early modern cognition, precisely because questions of structure or agency remain paramount in this latest line of investigation.

Distributed cognition models emerged in the late 1980s, with cognitive science research informing studies of human-computer interaction[17] and applied mathematics.[18] Jean Lave drew upon the latter body of work in developing a "social anthropology of cognition," which treated cognition as a social phenomenon: "The point is not so much that arrangements of knowledge in the head correspond in a complicated way to the social world outside the head, but that they are socially organized in such a fashion as to be indivisible."[19] Lave defined cognition, in a phrase oft since repeated by scholars sympathetic to the idea, as "distributed—stretched over, not divided among—mind, body, activity and culturally organized settings."[20] Lave's use of *mind* here is strategic, the comma being added between it and *body* purely to enable a point of differentiation based on the "venerable division of mind from body" that the notion of distribution is intended to undermine. Yet it is worth noting from the outset here that a key problem presents itself in trying to explain an embodied mind without recourse to an inherited language of mind *and* body.[21] The lurking danger is that terminological slippage can spill over into attribution error. This is one problem addressed by Ed Hutchins in the first sustained description of a distributed cognition model in *Cognition in the Wild* in 1995.[22] In earlier work, Hutchins had been modeling distributed cognition from detailed observations of work practices in aircraft cockpits[23] and in marine navigation,[24] with his anthropological background having equipped him to view the work of each individual in terms of collective procedures or mechanisms. His encounter with cognitive science in these studies brought Hutchins to the realization that while a fundamental tenet of this science remains correct—cognition can be defined as "computation realized through the creation, transformation and propagation of representational states"[25]—cognitive scientists had long since failed to correct their initial error of taking computation to be an abstract property of an individual mind, thus ignoring the social domains in which the individual functions on a daily basis. This focus on the individual had led cognitive science to reproduce what Hutchins sees as the key error in classicist appropriations of the Cartesian *cogito*: "When one commits to the notion that all intelligence is inside the inside/outside boundary, one is forced to cram inside everything that is required to produce the observed behaviours. Much of cognitive science is an attribution problem."[26] *Cogito ergo sum*; ergo, the sum of all that "I am" is then misattributed as the sum of all that "I think," having taken as read the foundational proposition of the dualist: my body cannot be trusted.

For Hutchins, then, rather than an abstract property of a mind, cognition is described as a concrete property of the system in which representations can be created, transformed, and propagated. He establishes a framework for conducting analyses of operations involving any number of people in terms of the cognitive properties of the group or, to be more precise, the *team*. We noted Hutchins's own earlier work on aviation and marine navigation; other studies by scholars working on distributed cognition have since looked at air traffic control towers,[27] call centers,[28] dairy farm apprentices,[29] hospital radiology departments,[30] joint design teams,[31] rescue command control centers,[32] and software programmers,[33] to nominate a few examples. To this list, we can add early modern theatre companies, thanks to Tribble's pioneering work in this area. This focus on distributed cognition within team activities ensures that the model developed by Hutchins has from the beginning resisted recapitulating the constraint-based explanations of the relationship between cognition and social context toward which social psychology has gravitated. It is not simply the case that an individual creates, transforms, or propagates representational states "with reference to" or "within constraints imposed by" a team and its operational focus; rather, these processes are described as taking place within the operations of the team—individuals' cognitive processes are not constrained nor are they even simply influenced by the context; the unit of analysis is the context itself, across which cognition is fully distributed. Hutchins sets out to explain how cognition *qua* computation is distributed throughout a culture and how it evolves over time as the cultural elements of any activity (tools, manuals, standard operations, hierarchies and protocols of control, and so on) are altered in even the most minute fashion by the operations of each team and the solutions they find to minor problems. In this way, Hutchins concludes, cognition is "a fundamentally cultural process."[34] Rather than being constrained by social contexts, cognition is the modus operandi of culture writ large.

Tribble is, admittedly, not the first scholar to focus on early modern theatre companies. Out of the tumult of the "subversion" versus "containment" debates that shaped much of the work being done under the banner of Cultural Materialism and New Historicism in the 1980s and early 1990s, a generation of scholars has drawn on methods from cultural history in their attempts to paint a picture of Shakespeare's theatre and its world with far greater detail than earlier generations of scholars might have thought possible or, for that matter, necessary. One good thing to come out of the "subversion" versus "containment" wars was a realization that, even in the attempt to overturn the hegemony of text-centered New Criticism and in a move intended to avoid a return to an older form of historicism, the historicists of the 1980s ended up reiterating a form of study based on the Elizabethan world view. The only real difference was the debate over top-down or bottom-up mechanisms sustaining this worldview. Out of this debate emerged the generation of scholars who were interested in neither possibility—the prospect that Shakespeare was either contained by the monarch's boundless power, or that his subversive theatre radically refigured Elizabethan power structures—but wanted to arrive at a fuller understanding of the cultural place of the theatre and the many layers of institutions and people attenuating any relationship between playwright and crown. Tribble's contribution to this recent turn to cultural histories of Shakespeare's theatre is to bring into play a model that had been explicitly designed to explain group processes without recourse to the intentions or the will of individual agents.

The particular value of the distributed cognition approach as applied by Tribble is in the status it prescribes to the textual and material residues of the early modern theatre company as "cognitive artifacts," the tools and "smart structures"[35] through which cognitive powers are exercised and onto which excessive cognitive complexity can be viably "offloaded."[36] Where the marine navigators studied by Hutchins had their charts, GPS systems, electronics, control room layout, and even "the social structures of shipboard life,"[37] the material residue from the practices of the early modern theatre companies includes parts, plots, playbooks, the evidence of the architecture of the playhouses, documents relating to boy players and apprenticeships in theatre companies, and of course what the plays themselves tell us about acting, memory, and other such matters. Importantly, in treating these material residues as cognitive artefacts, Tribble does not simply see them as evidence for what early modern actors did (as if texts or objects are only the end-products of cognitive processes and resulting practices); rather, they are treated as the smart structures through which the actors and audiences exercised cognitive processes in the playhouses. If there is any irony here, it is to be found in Tribble's use of the word *constraint* at just this moment in her description of the relationship between cognition and its artefacts: "human beings create smart structures predicated on a system of constraints that, paradoxically, enable the execution of complex tasks."[38] Unlike the constraint models of other versions of social psychology, though, Tribble uses the term here to mark the paradox: what some characterize as the constraints on cognition, Hutchins and his adherents view as a proliferation of extended structures enabling or indeed constituting cognition. Importantly, in the focus on "smart structures," the distributed cognition model holds that these structures are embedded in the materials and processes peculiar to each group process—this is, in a sense, what makes them "smart"—rather than imagining the structure itself as being imposed upon, and defining or shaping from without, the localised operations of the group activity. This is a point to which I shall return in a moment, as it speaks to a potential tension within application of the distributed cognition model to historical phenomena or activities.

To arrive at this point of tension, it is first necessary to understand how a cultural history of early modern cognition can overcome the tyranny of historical distance such that it can be said to present a relatively accurate account of cognition in The Globe in the first years of the seventeenth century. To what extent, we might always cautiously ask, is the explanation of a historical process oriented principally towards the explanatory model or toward the records through which we make sense of the past? This is no simple equation, wherein history should for example always be allowed to speak for itself. If history speaks, do we nevertheless miss something in translation? When Hamlet asks if it is "nobler in the mind" to suffer a range of physical torments produced by misfortune (*Hamlet* 3.1.59),[39] do we hear the word *mind* in a suitably pre-Cartesian frame?[40] Our explanatory models become most necessary when we seek to explain that which is most foreign to us, and the past is, to be sure, a foreign place. In *Cognition in the Globe*, Tribble admits to having stumbled on her previous attempt to explain how the preparation and performance of early modern plays might be understood in terms of distributed cognition: the particular stumbling block was the current debates over whether Shakespeare's company worked with a two-door stage (left and right stage only) or a three-door configuration (with an added door at center stage).[41] In "Distributing Cognition in the Globe," Tribble weighed up the two sides of the debate: on the one hand, Andrew Gurr and Mariko Ichikawa followed the work of Bernard Beckerman in arguing that actors typically entered through one door, crossed the stage during the scene, and exited via another, yet their model relied also on the flexibility of a central third door; on the other hand, Tim Fitzpatrick proposed a two-door model in which one door represented an offstage space that was "further inwards" while the other was "further outwards," and actors calculated entries and exits via this spatial logic.[42]

In this 2005 article, Tribble determined that Fitzpatrick's model would have been "risky at best" as a convention for governing stage movement: "The more thinking that can be off-loaded onto the environment, the more mental energy remains available for those tasks that are primarily internal (memory for the spoken lines, for instance)."[43] The risk, as it presented itself to Tribble in 2005, was that Fitzpatrick's model did not account for varying degrees of difference in "fictional space" from play to play and even from scene to scene.[44] The mental energy required to "map fictional onto geographic space" within the work space of the stage would have reduced the amount of energy available to remember lines, among other things.[45] Yet in *Cognition in the Globe*, Tribble reconsidered this conclusion, observing instead that a meaningful convention could most likely have reduced the cognitive demands on both actor and audience. In the six years between publications, Tribble had given more thought to the principle of "cognitive thrift" that is expressed earlier in terms of the minimization of loads on internal mental energy but which by 2011 is understood as both minimization of internal cognitive load (off-loading) and the establishment of conventions, skills, and smart structures to ensure load is reduced to begin with (underload).[46] The question of two or three doors thus becomes one of whether or not some of the spatial thought required by a play can already be done by the doors themselves, so to speak. A convention that locks the actors into entering and exiting via the same door actually increases the interpretive work required to make sense of the spatial logic of the play, which includes the orientation of actors *on stage*. Tribble cites the example of Act 2, Scene 2 of *King Lear*, in which Oswald's opening query about "this house" (TLN 1076) clearly establishes that one door will be designated as Gloucester's house for this scene.[47] All other references to that location should also therefore be oriented by the actor toward the same door. A convention dictating that one door was usually more inward would also enable actors to not have to think about the spatial orientation of "this house," as this spatial logic has already been thought through for them in advance.

One way to think about Tribble's conclusions in the 2005 article, I suggest, is to imagine at their core the persistence of the inside/outside binary against which a distributed cognition model is directed. Allow me to explain. The idea that thought can be *off-loaded* onto X relies on the logical proposition that the normal site in which thought *is* loaded *is not* X; if thought is *off*-loaded onto the environment, then, the normal location for cognitive loading is internal, set apart from the external environment. The process is thus configured as being situated in a *moment* of cognition—a here-and-now of the thinking activity of the individual—with a brief opportunity to reduce load presented as the offsetting of the cognitive load from the internal mind to the external environment. In rejecting Fitzpatrick's model in 2005, Tribble slips into this language of inside and outside—memory, for example being identified as an internal task: "tasks that are primarily internal (memory for the spoken lines, for instance)." By focusing in subsequent work on more of the technical apparatus or smart structures used by early modern actors in staging plays, Tribble's model, informed by a richer cultural history, moves toward an understanding of pre-Cartesian cognition as more fully distributed than her 2005 work had allowed. The textual residues we have from the early modern theatre do not simply give us insight into what actors did or thought; quite literally, these residues present actors' actions or thoughts in direct fashion. By extending that thought from the play text to the architecture of the stage, Fitzpatrick now provides Tribble with the necessary counterpoint to the binary of inside and outside: the mind of an early modern actor need not have been structured to mirror the architecture of a stage, as if this is required to enable the actor's thoughts to be perfectly suited to the performance environment; rather, stage convention thinks the required spatial logic for the actors.

Another way to view the advantage of using cultural history for this project is to consider the difference between cognition as a momentary activity—a here-and-now as I described it a moment ago—and cognition as taking place across apparatuses and smart structures (spatially extended and therefore more viably conceived as process rather than as momentary activity). When thrift is applied to a momentary activity, the goal would seem to be optimum efficiency in the use of all related materials; yet cultural histories might be useful in reminding us that in the purpose-built playhouses like the Globe, and in the "smart structures" developed for use by the actors, conventions might have been developed according to principles of efficiency in a given set of circumstances but might not have translated as the optimal use of the Globe or other theatres like it. The history of the technical evolution of a more recent invention—the keyboard layout on which the present chapter was typed—informs us that thrift is oftentimes superseded by a range of other factors. The

QWERTY layout was developed by Christopher Sholes and Amos Densmore in the early 1870s to minimize jamming in manual typewriters. While the ball-head electric typewriter later solved the jamming problem and the peripheral devices used in personal computing (keyboard, monitor, and mouse) remove the typewriting technology altogether from the word processing activity, the keyboard continues in standard formats to adhere largely to a model that is 140 years old. Paul David observes that during the last decades of the nineteenth century, the development of touch typing, along with the growth of typing as a socially and economically significant form of labor, caused QWERTY to be "locked in" as a standard layout.[48] With too much industrial and social capital invested in the QWERTY layout, it will take more than the arguments about the efficiency of other layouts for QWERTY to be supplanted. Donald Norman puts the issue in sharp perspective: "We are committed to [QWERTY] even though it was designed to satisfy constraints that no longer apply. . . . The severe constraints of existing practice prevent change."[49] It is on this question—of the technical evolution of the smart structures on which the model of distributed cognition relies—that a model of constraint is appropriately brought into play. As Norman reminds us, the constraints of existing practice can be "severe," heavily outweighing any developments based on efficiency alone. Similarly, we should in no way imagine that the purpose-built playhouses radically overturned received theatrical practice in a blink of the eye, historically speaking.

Prior to the advent of large purpose-built playhouses, the life of the acting company was a highly transitory one, which means that the conventions developed for staging plays will have needed to be highly transportable: a play performed in a private hall in one location could just as easily have been performed in a town square in another. This issue of transportability did not disappear when, in 1567, The Red Lion was constructed for the purpose of housing public performances of plays, nor had it ceased to be an issue when James Burbage and associates established the Theatre in Shoreditch nine years later.[50] Tours to provincial locations and to continental Europe, court performances, and paid private performances remained very much a part of the stock in trade of the acting companies throughout the era of the purpose-built playhouses. Rather than imagining that the smart structures adopted by actors were suited to the most optimal use of any single site, then, we might instead consider the prospect of there being adaptability built in to the constraints of existing practice. In Chapter 1 of *Playwright, Space and Place in Early Modern Performance: Shakespeare and Company*, Fitzpatrick's first example of a two-door exit is in the parting of the ways by the Queen and Suffolk at the end of Act 3, Scene 2 of *2 Henry VI*:

SUFFOLK. This way fall I to death
QUEEN. This way for me. Exeunt [severally][51]

The text itself cues the departure by two actors through two different doorways. Importantly, this is not an example that signifies anything of consequence for any specific theatrical space: though it was published in Quarto in 1594, the play was certainly well known by 1592, when Robert Greene's attack on Shakespeare as an "upstart crow" was published with its reference to his "tiger's heart wrapped in a player's hide" in a straight parody of York's "tiger's heart wrapped in a woman's hide."[52] Fitzpatrick's first example of a two-door exit was written not for the Globe, nor was it even written necessarily for The Theatre. *2 Henry VI* was written in the period loosely referred to as Shakespeare's "lost years," which we know simply by virtue of the fact that it is the play parodied by Greene in what is conventionally considered to be the first definitive reference to Shakespeare as a London playwright, so we do not know for which company the play was written nor in which venue it was first performed. Accordingly, the example of a textually cued two-door exit in this play signifies that Shakespeare, for one, was writing for a two-door system from very early in his career, long before he became more settled at the Theatre with The Lord Chamberlain's Men from 1594 and before this company established the Globe for themselves in 1599.

Fitzpatrick's later examples of "in" and "out" being deployed in Elizabethan play texts and stage directions also include a number of prominent early non-Shakespearean examples, such as Kyd and Marlowe, to show that this convention predates Shakespeare.[53] The model is thus not to be limited to a theory of stage management at the Globe; it is an account of a practice that must have evolved among the itinerant acting companies whose plays needed to be able to be staged in potentially any space and then adapted for performance in situ. Despite the title of her 2005 article and her more recent book, Tribble's cultural history of early modern cognition among players and their audiences is also not restricted to the Globe. This will to consider the emergence of these smart structures throughout different locations and across a stretch of historical time is necessary to offset the potential tension to which I referred earlier and which I will now outline more directly. The method used by Tribble to offset historical distance—as indeed by many who have embraced the cultural turn in Shakespeare Studies in the wake of *Renaissance Self-Fashioning*—is a "thick description of a particular social and material moment" in examining early modern stage practices.[54] As Roger Chartier argued in the first wave of responses to the Geertzian turn in historical studies, perhaps most notably in Robert Darnton's *The Great Cat Massacre and Other Episodes in French Cultural History* of 1984, the risk of thick description rests in its capacity to conflate actions or practices with the objects of *reading*: texts.[55] Chartier demands instead that the cultural historian, of all people, should be mindful of the subtle differences in form and purpose that different types of objects may possess within their culture of origin:

> Metaphorical use of terms like "text" or "reading" is always risky, and it is even more so when the only object under anthropological investigation is a *written* text. Not only does it obliterate the ways of speaking or acting that gave the tale or the rite as much significance as its literal meaning (or even more); above all, a real text with a status of its own stands between the observer and this oral or festive supposed "text."[56]

Thick description is a method for extrapolating broader cultural structures from any observed phenomenon by virtue of gathering a wide array of descriptive details about that phenomenon and, as a result, relies on the ability of the observer to convert observation into text even if the cultural object does not exist in written form. For the cultural historian, the bulk of evidence constituting the encounter with another culture is already textual, and the risk will be that the text will be read primarily for its capacity to "describe" the past rather than to be a product of practices peculiar to a "particular social and material moment," as Tribble puts it. Alan Liu has characterised thick description and the entire cultural turn within History and Literature as being reliant on these descriptions parsing as fragments: "Such rhetoric clearly confesses incompletion, but, at last, also the unmistakable sign of synecdoche."[57]

In thick description, then, there remains the ever present risk of subsuming the residues of the past to the heuristic objectives and rhetorical strategies of the observer, particularly where the method is being employed precisely for its ability to provide access to the structures that are presumed to have shaped the object from without. Tribble's thick descriptions avoid such pitfalls because the objects are being viewed as smart structures, in and of themselves. In this sense, at least, we might state that the distributed cognition model gives to cultural history an invaluable corrective to some of the potential risks presented by methods inherited from those who looked to Geertz for an understanding of how "cultures" might be interpreted. It is worth remembering, moreover, that the distributed cognition model *pace* Hutchins should also give to cultural history a nuanced understanding of how smart structures evolve over time, adding an appreciation of historical process, as it were, rather than simply an awareness of historical distance. Hutchins gives an account of the gradual technical evolution of systems and smart structures in his model: each moment of distributed cognition is also shaped by a history of the "development of the practitioners" and the even longer history of the "development of the practice."[58] Yet the studies of phenomena which constitute the bulk of the field of writings on distributed cognition tend to focus principally on an activity as it has been observed in action, thereby leaving an account of the long-term development of the practice to a brief footnote at best. Gavriel Salomon's own contribution to his collection, *Distributed Cognitions*, is suspicious of the potential for a focus on systems to overlook the role of the individual members of the team on systemic change: "change is brought about by affecting the cognitions or behaviours of one or another member . . . whereby a quasi-stationery equilibrium is upset, thus requiring all other components to adjust."[59] Salomon contends that the team-oriented focus of the bulk of studies of distributed cognition leads to the same heuristic bias that we have identified with Geertzian thick description: team studies provide "*a snapshot frozen in time* that captures but half of the process" (emphasis in the original).

I am not convinced that the distributed cognition model leads invariably to cross-sectional or "snapshot" analyses: for mine, this is a quibble with some applications of the model; not a consequence of the model itself. The explanation that Hutchins gives of technical evolution at the heart of the distributed cognition model clearly articulates a role for the development of the practitioner. If there is a question left unanswered by Hutchins, at least in his 1995 book, it concerns not the individual per se but the body of the individual. We have noted that in seeking to correct the attribution error underlining and undermining cognitive science, based on the Cartesian inside/outside boundary, Hutchins shifts the site of cognition to the outside. What is never entirely clear in this shift in *Cognition in the Wild* is whether the body of the human agent is seen as the locus of the boundary or a component of the outside. Around the same time, John Haugeland was promoting the conception of mind as "not incidentally but *intimately* embodied and *intimately* embedded in its world,"[60] but as Andy Clark points out, the nature of the intimacy invoked by Haugeland remained a "famously slippery beast" with a drift toward philosophical externalism—an appeal, for example, to the role of the actual road in orienting an agent's knowledge of the way to San Jose—continuing to ground studies of a mind embodied and embedded for most of the next decade.[61] In more recent work, Hutchins has incorporated ideas related to "somatic anchoring" into the distributed cognition model, to explain how the body participates in cognition: gestures play a key role in helping us to "see" for ourselves the spatial and motile logic of complex or abstract concepts and relations.[62] To extrapolate further along the same lines, contra Haugeland's famous example, the road may well play a key role in orienting us toward San Jose, for a particular kind of externalism, but the body must also "know" what to do with a road. Such considerations must of course come into focus with sharp relief when we grapple with what distributed cognition might look like in a world that preceded the dualist separation of mind from the body that cannot be trusted. How, in short, might we come to "know" what early modern bodies knew? While we cannot observe early modern bodies using gestures to aid them in anchoring concepts, as Hutchins does with bodies in the modern workplace, we can, to some extent, allow the early moderns to speak for themselves, including their gestures.

Tribble devotes a significant portion of *Cognition in the Globe* to the early writings on gesture by John Bulwer.[63] Bulwer's *Chirologia* and *Chironomia* (published together in 1644) document the belief that gestures are the "naturall language" of the hands and head.[64] Tribble notes that Bulwer's writings and, in particular, his illustrations of the "Alphabet of naturall Gestures of the Fingers," have in the past been widely misrepresented by scholars of the early modern period as prescriptive; they are, instead, a catalogue of the natural bodily dispositions to which gestures were thought to belong.[65] An aspiring actor would have needed, of course, to be able to discipline this natural language but the language itself is viewed by the early moderns as wholly consonant with the admixture of body with mind. Accordingly, as Tribble points out, early modern gestural language is more than simply embodied; it is a version of an extended or distributed mind hypothesis:

> The movements of the hand are part of a *techne*, a distributed system harnessing the hand-brain-language nexus, extended social networks, treatises, institutions . . . and embodied expertise, including experience in rhetorical invention and, in the case of actors, ability to parse written language for its action.[66]

Importantly, once the focus switches to the actor and his body, the potential for examining an evolutionary process *in situ* also becomes apparent. The analysis of early modern approaches to gesture, action, and accent, includes detailed examination of evidence from early modern play texts, including the formative early English play *Gorboduc* (1561) and *The Spanish Tragedy* (1588). Analysis of the former leads Tribble to conclude that this play was written without recourse to a "*gesture-potentiated* language" that is evident, for example, in Bulwer's later treatises.[67] *The Spanish Tragedy*, on the other hand, marks innovation of embedding a language that yokes action (gesture) to words into the speeches of Hieronomo.[68]

If we might wonder how the language of the Elizabethan players manages to evolve in the space of less than three decades in such dramatic fashion, Tribble's chapter on "Enskillment in the Early Modern Theatre" provides some answers: in short, the rise of the apprenticeship system brought with it a range of internal practices within the theatre that enables boy actors to learn on the job, so to speak.[69] The language that maps gestures to words becomes wholly internalised, fitting play texts more perfectly to the environment in which bodies and voices will be used to convey the text. Cognition in the early modern theatre thus evolves not only by dint of principles of thrift or efficiency but by the natural evolution of the stage to adapt to a language of the early modern body-mind, and by adaptation of that language to the shifting demands of the theatre. Distributed cognition, suitably expanded to cover the work of the cultural historian, is thus a valuable model for understanding the early modern theatre given the proviso that the bodies of actors are not seen as the limit point or threshold across which philosophical conundrums of inside or outside need to be thought out. Instead, these bodies are present within the system and, particularly in an era in which the Cartesian split had not yet taken place, they are points of reference for any representations that might otherwise be called the work of cognition.

If I could argue for one further adjustment to the distributed cognition model as applied to the early modern stage—or, indeed, in general—it would be in returning once more to the role of individual change in systemic change. Here I am drawn to considering the company personnel or perhaps the patron, the audience, or other key persons as catalyzing factors, not even necessarily as cognizing agents for change within a system of distributed cognition. We know, for example, that the Lord Chamberlain's Men were shaken in 1599 when the leading clown William Kemp departed on what we believe were less than amicable terms: Kemp's departure, as David Wiles has argued, may well have represented "a significant jolt on the graph of theatrical change."[70] A single actor leaves a company, and the ripple effect created by this departure impacts on the company's repertory, the way their playwright shapes plays for comic effect thereafter, the tradition of the clown as a stock figure, and even the types of entertainments offered at the theatres, with the decreasing importance of jigs during Robert Armin's time. Systems evolve, but they do not evolve systematically. The slightest change in personnel or even in relations between existing personnel can create flow-on effects that alter an entire system at the level of what we would recognise after the fact as an evolutionary shift. Perhaps a future goal of studies of distributed cognition within early modern theatre companies might well become to marry the messiness that emerges from detailed studies of company personnel, repertories, patrons, and audiences to an approach that guides us to think of change in epochal, systematic terms. One of the gains that can be made by looking for the potential impact of individuals on system change, particularly within our cultural histories of the early modern theatre, is that it will further inure us against any return to the subversion or containment models that have held sway for the better part of three decades in Shakespeare Studies. Shakespeare need not be imagined as an agent for subversion *or* an Elizabethan puppet; instead, he can be returned to a vibrant and active company that went about the business of staging plays, new and old, oftentimes within the severe constraints of existing practice and sometimes as innovators, but always as practitioners governed by the coupling of thought to action in the theatre.

NOTES

1. Greenblatt's examination of Geertz's *The Interpretation of Cultures* framed his foundational *Renaissance Self-Fashioning*; his explanation of the containment model that arises from this approach was outlined in the essay "Invisible Bullets," which was published four times between 1981 and 1988. See Greenblatt, *Renaissance Self-Fashioning: From More to Shakespeare* (Chicago: University of Chicago Press, 1980): 3–6; Greenblatt, "Invisible Bullets," *Glyph* 8 (1981): 40–61; and for examinations of the "containment" model from differing approaches to the debate, see Hugh Grady, "Containment, Subversion—and Postmodernism," *Textual Practice* 7.1 (1993): 31–49; Steven Mullaney, "After the New Historicism," in *Alternative Shakespeares,* Vol. 2, ed. Terence Hawkes (London: Routledge, 1996): 17–37; and Jonathan Gil Harris, "Historicizing Greenblatt's Containment: The Cold War, Functionalism, and the Origins of Social Pathology," in *Critical Self-Fashioning: Stephen Greenblatt and the New Historicism*, ed. Jürgen Pieters (New York and Frankfurt: Peter Lang, 1999): 150–173.
2. See Jonathan Dollimore, "Introduction: Shakespeare, Cultural Materialism and the New Historicism," in *Political Shakespeare: Essays in Cultural Materialism*, ed. Jonathan Dollimore and Alan Sinfield (Manchester: Manchester University Press, 1985): 2–17. A revised version of Greenblatt's "Invisible Bullets" is, incidentally, the first essay included in this collection.
3. Evelyn Tribble, "Distributing Cognition in the Globe," *Shakespeare Quarterly* 56.2 (2005): 135–155.
4. Evelyn B. Tribble, *Cognition in the Globe: Attention and Memory in Shakespeare's Theatre* (New York: Palgrave Macmillan, 2011).
5. W.H.R. Rivers, "Sociology and Psychology," *The Sociological Review* 9 (1916): 1–13; esp. 12.
6. Peggy A. Thoits, "Social Psychology: The Interplay between Sociology and Psychology," *Social Forces* 73.4 (1995): 1231–1243; esp. 1231.
7. Greenblatt, *Renaissance*, 7–9.
8. Clifford Geertz, qtd. in Greenblatt, *Renaissance*, 3.
9. Stephen Greenblatt, *Shakespearean Negotiations: The Circulation of Social Energy in Renaissance England* (Oxford: Clarendon Press, 1988), 30.
10. Linda Charnes, *Notorious Identity: Materializing the Subject in Shakespeare* (Cambridge, Massachusetts: Harvard University Press, 1993), 18.
11. Stephen Greenblatt, *Hamlet in Purgatory* (Princeton, New Jersey: Princeton University Press, 2002); *Will in the World: How Shakespeare Became Shakespeare* (New York: W. W. Norton, 2005); *Shakespeare's Freedom* (Chicago: University of Chicago Press, 2010). I offer an examination of such tendencies across Greenblatt's career in "Shakespeare's Gifts: Self-Fashioning, Authorising, Stephen Greenblatt," in *"Rapt in Secret Studies": Emerging Shakespeares*, ed. Darryl Chalk and Laurie Johnson (Newcastle upon Tyne: Cambridge Scholars Publishing, 2010): 329–346.
12. Philippa Kelly, ed. *The Touch of the Real: Essays in Early Modern Culture, in Honour of Stephen Greenblatt* (Crawley: University of Western Australia Press, 2002).
13. Stephen Greenblatt, "The Touch of the Real," *Representations* 59 (1997): 14–29.
14. Patricia Parker, *Shakespeare from the Margins: Shakespeare, Culture, Context* (Chicago: University of Chicago Press, 1996), 11.
15. Parker, 55.

16. Stephen Greenblatt, *Shakespeare's Freedom* (Chicago: University of Chicago Press, 2010), 1.
17. Liam J. Bannon, "A Pilgrim's Progress: From Cognitive Science to Cooperative Design," *AI & Society* 4.4 (1990): 259–275.
18. Jean Lave, *Cognition in Practice: Mind, Mathematics, and Culture in Everyday Life* (Cambridge: Cambridge University Press, 1988).
19. Lave, 1.
20. Lave, 1.
21. Lave, 180.
22. Ed Hutchins, *Cognition in the Wild* (Cambridge, Massachusetts: MIT Press, 1995).
23. Ed Hutchins and Don Norman, "Distributed Cognition in Aviation: A Concept Paper for NASA," Contract No. NCC 2–591 (San Diego: Department of Cognitive Science, University of California, 1988).
24. Edwin Hutchins, "The Technology of Team Navigation," in *Intellectual Teamwork*, ed. Jolene Galagher, Robert E. Kraut, and Carmen Egido (Hillsdale, New Jersey: Lawrence Erlbaum Associates, 1990): 191–220.
25. Hutchins, *Cognition*, 49.
26. Hutchins, *Cognition*, 355.
27. Christine A. Halverson, "Distributed Cognition as a Theoretical Framework for HCI: Don't Throw the Baby Out with the Bathwater—The Importance of the Cursor in Air Traffic Control," Report 9403 (San Diego: Department of Cognitive Science, University of California, 1994).
28. Mark S. Ackerman and Christine A. Halverson, "Considering an Organization's Memory," in *Proceedings of the ACM Conference on Computer Supported Cooperative Work* (CSCW '98) (New York: ACM, 1998): 39–48.
29. Cristina Grasseni, "Skilled Vision: An Apprenticeship in Breeding Aesthetics," *Social Anthropology* 12.1 (2004): 41–55.
30. Yvonne Rogers and Judi Ellis, "Distributed Cognition: An Alternative Framework for Analysing and Explaining Collaborative Working," *Journal of Information Technology* 9.2 (1994): 119–128. Rogers and Ellis also consider the example of a workplace in which shared networking arrangements regulate computer-based work activities.
31. Mark Perry and Duncan Sanderson, "Coordinating Joint Design Work: The Role of Communication and Artefacts," *Design Studies* 19.3 (1998): 273–288.
32. Christer Garbis and Yvonne Waern, "Team Coordination and Communication in a Rescue Command Staff—the Role of Public Representations," *Le Travail Humain* 62.3 (1999): 273–291.
33. Nick V. Flor and Edwin L. Hutchins, "Analyzing Distributed Cognition in Software Teams: A Case Study of Collaborative Programming during Adaptive Software Maintenance," in *Empirical Studies of Programmers: Fourth Workshop*, ed. Jürgen Koenemann-Belliveau, Thomas G. Moher, and Scott P. Robertson (Norwood, New Jersey: Ablex Publishing, 1991): 36–64.
34. Hutchins, *Cognition*, 374.
35. Tribble, *Cognition*, 5.
36. Tribble, *Cognition*, 8.
37. Tribble, *Cognition*, 5.
38. Tribble, *Cognition*, 5.
39. My references to Shakespeare plays are from John Jowett, William Montgomery, Gary Taylor, and Stanley Wells, eds., *The Oxford Shakespeare: The Complete Works,* 2nd ed. (Oxford: Oxford University Press, 2005).
40. For a discussion of Hamlet's reference to "mind" in terms of contemporary meanings of mind, see Laurence Johnson, "'Nobler in the Mind':

The Emergence of Early Modern Anxiety," *AUMLA* Special Issue (2009): 141–156.

41. Tribble, *Cognition*, 33–41.
42. Evelyn Tribble, "Distributing Cognition in the Globe," *Shakespeare Quarterly* 56.2 (2005): 135–155, esp. 143.
43. Tribble, "Distributing," 144.
44. Tribble, "Distributing," 144.
45. Tribble, "Distributing," 143.
46. Tribble, *Cognition*, 20.
47. Tribble, *Cognition*, 38.
48. Paul A. David, "Clio and the Economics of QWERTY." *The American Economic Review* 75.2 (1985): 332–337; esp. 334.
49. Donald A. Norman, *The Design of Everyday Things* (New York: Doubleday, 1988), 147, 150.
50. Herbert Berry disputes the long-standing convention of treating the construction of the theatre as the origin of the purpose-built playhouses around London in his account of the history of the Red Lion, "The First Public Playhouses, Especially the Red Lion," *Shakespeare Quarterly* 40.2 (1989): 133–148.
51. Tim Fitzpatrick, *Playwright, Space and Place in Early Modern Performance: Shakespeare and Company* (Farnham and Burlington: Ashgate, 2011), 11.
52. Peter Ackroyd, *Shakespeare: The Biography* (London: Vintage, 2006), 176.
53. Fitzpatrick, 199–208.
54. The method derived by New Historicists and cultural historians from Clifford Geertz. Tribble, *Cognition*, 8.
55. Roger Chartier, "Texts, Symbols, and Frenchness," *The Journal of Modern History* 57.4 (1985): 682–695.
56. Chartier, 685.
57. Alan Liu, "Local Transcendence: Cultural Criticism, Postmodernism, and the Romanticism of Detail," *Representations* 32 (1990): 75–113; esp. 92.
58. Hutchins, *Cognition*, 372.
59. Gavriel Salomon, "No Distribution Without Individuals' Cognition: A Dynamic Interactional View," in *Distributed Cognitions: Psychological and Educational Considerations*, ed. Gavriel Salomon (Cambridge: Cambridge University Press, 1996): 111–138; esp. 119–120.
60. Haugeland's idea of embodied and embedded minds had been in circulation since the early 1990s. The phrase here is cited from Haugeland, "Mind Embodied and Embedded," in *Having Thought: Essays in the Metaphysics of Mind*, ed. John Haugeland (Cambridge, Massachusetts: Harvard University Press, 1998): 207–240; esp. 237.
61. Andy Clark, "Pressing the Flesh: A Tension in the Study of the Embodied, Embedded Mind," *Philosophy and Phenomenological Research* 76.1 (2008): 37–59; esp. 37–39.
62. Edwin Hutchins, "Imagining the Cognitive Life of Things," *The Cognitive Life of Things: Recasting the Boundaries of the Mind*, Symposium. McDonald Institute for Archaeological Research, Cambridge, April 2006. See also Hutchins, "Enaction, Imagination, and Insight," in *Enaction: Toward a New Paradigm for Cognitive Science*, ed. John Robert Stewart, Olivier Gapenne, and Ezequiel A. Di Paolo (Cambridge, Massachusetts: Massachusetts Institute of Technology, 2010): 425–450; esp. 442–443.
63. Tribble, *Cognition*, 90–102.
64. Tribble, *Cognition*, 99.
65. Tribble, *Cognition*, 99–100.
66. Tribble, *Cognition*, 101.

67. Tribble, *Cognition*, 105.
68. Tribble, *Cognition*, 107.
69. Tribble, *Cognition*, 111–150.
70. David Wiles, *Shakespeare's Clown: Actor and Text in the Elizabethan Playhouse* (Cambridge: Cambridge University Press, 2005), xi.

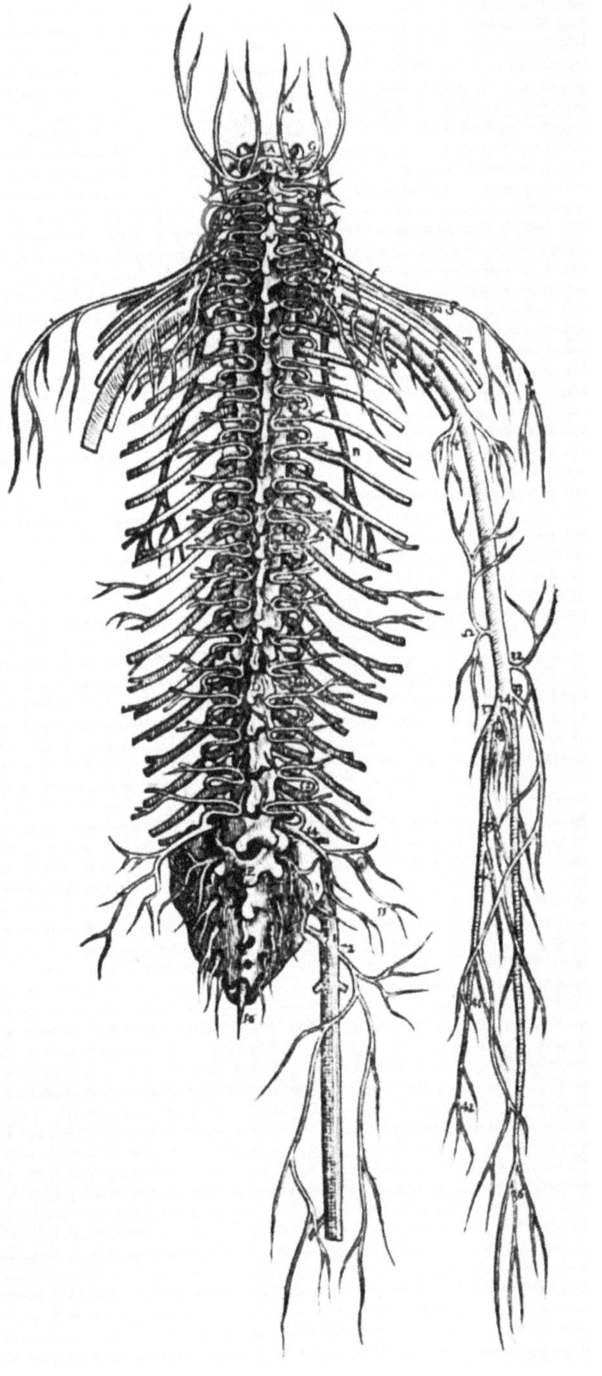

Figure 9 "Nervous System Plate," from Vesalius, *De humani corporis fabrica* (1543), 332 (Liber IV). Image A027043, History of Medicine Collection, Courtesy U.S. National Library of Medicine.

12 The Belly-Mind Relationship in Early Modern Culture

Digestion, Ventriloquism, and the Second Brain

Jan Purnis

In the past few decades, cognitive scientists have been investigating the mind and its processes, tending to privilege the brain as the locus of the mind even when acknowledging the importance of the brain-body relationship in mental and emotional experience. At the same time, neurobiologists like Michael Gershon, M.D., who work more specifically in the field of neurogastroenterology, have proposed that there is a "second brain" in the gut, noting that there are more nerve cells in the esophagus, stomach, and large intestine than in the rest of the peripheral nervous system, and that this enteric nervous system (ENS) in the gastrointestinal tract contains every category of neurotransmitter found in the brain. In *The Second Brain*, his 1998 book on nervous disorders of the stomach and intestine, Gershon suggests that the "ugly gut is more intellectual than the heart and may have a greater capacity for 'feeling,'" that it is an independent site of neural integration and processing, and that it may have its own psychoneuroses.[1] This notion of a brain in the gut has profound implications for thinking about the body-mind relationship, and it has the potential to challenge the cerebrocentrism that aligns the mind with the brain in the head. Arguing for the usefulness of neuroscience to feminist theories of the body and drawing upon early psychoanalysis, Elizabeth A. Wilson urges a vigorous reconceptualization of the psyche/soma relationship in light of recent research on the gut. Examining modern eating disorders and depression, Wilson asserts that not only is the gut in "natural alliance with psychological events" but "there is something about the gut that makes it a particularly potent psychological organ" and yet "the psychodynamics of this part of the nervous body remain understudied."[2] Also understudied by medical researchers and theorists are historical perspectives on the gut from before the nineteenth century. While Gershon observes that a "renaissance of the gut is under way," he adds that the "status of our knowledge of the enteric nervous system has been, until recently, positively medieval. Medieval ignorance, however, yielded once before to the Renaissance and the Renaissance led eventually to the Enlightenment."[3] Gershon's temporal metaphor for scientific progress misses a key point: that the so-called Enlightenment in many ways eclipsed earlier beliefs about the gut's psychophysiological importance, which are highly relevant to recent "discoveries." Researchers participating in this scientific "renaissance of the gut" might thus do well to consider Renaissance understandings of the gut, as might researchers in the field of cognitive science.

In what follows, I offer a historical context for theorization about the body-mind relationship through an exploration of early modern understandings of the gut's function and its relationship to the brain and to the mind. While the word "gut" is regularly used to refer to the alimentary canal in its entirety, it can also refer more specifically to the intestine or bowel. Because of my interest in the early modern period, I focus not solely on the intestine and stomach but on the abdominal region as a whole in order to better reflect early modern medical interpretations of the digestive process and the organs involved in it. As I will demonstrate, for Shakespeare and his contemporaries, the idea that the gut is strongly allied to psychological experience and is itself a "particularly potent psychological organ," as well as the idea that it shares similarities with the brain in the head, would be consistent with many of their medical and cultural views about the digestive organs contained in the belly.

In the Renaissance cultural and medical models of the self, the belly was one of the three regions into which the body was generally divided. Each region was ruled by one of the three principal parts: the brain in the head, the heart in the chest, and the liver in the belly. In the allied tripartite model of the soul, the rational, affective, and nutritive faculties of the soul had their seats in the brain, heart, and liver respectively. Noting that "[f]or most of us today, the affective and intellectual are both located in the brain," Scott Manning Stevens has highlighted that "seventeenth-century physiology was only beginning to question the function of the heart as a muscle and not as an organ of thought, or at least not the center of our emotional life"; since the categories of thought and emotion were not rigidly defined, "Donne can refer to his 'naked thinking Heart' in 'The Blossom,' just as Shakespeare's Prospero describes his 'beating mind."[4] In addition, Christianity brought to the heart "the perceptive abilities of the conscience": "Because the heart, enlightened by the Holy Spirit, could *perceive*, it was more than the receptor of the passions. The heart could on some level compete with the brain as a locus of perception through its ability to perceive things unseen."[5]

Stevens remarks that of the three principal organs, the liver was never a serious contender as "the source of the self";[6] but while not perhaps as obvious a contestant, the liver and other organs of the belly were often associated with functions similar to those he explains were associated with the heart. For instance, the liver was connected to emotions like love, anger, and cowardice, a connection descended from ancient times when, according to Richard Broxton Onians, the liver appears to have been regarded as "the inmost spring of the deeper emotions" as well as the seat of "consciousness not limited to love and anger."[7] Ancient belief also held that it was "chiefly" out of not only the heart but also the reins [kidneys/kidney region] that "the thoughtes and cogitations of the mynde, and all licentious lustes and dissolute desyres, do proceede and springe, which are not nor cannot lye hyd, or unespyed of God," which, Levinus Lemnius explains to his early modern readers, is why in the Bible David uses the example of the heart and reins to stress that God "perfectly searceth out and knoweth all thinges" even in "the most secret corners and innermost places" of the self.[8] In an example from Shakespeare's lifetime, in addressing the troops in Tilbury at the time of the Spanish Armada, Queen Elizabeth is famously reputed to have urged: "I know I have the body of a weak and feeble woman, but I have the heart and stomach of a king, and of a king of England too."[9] While the stomach and heart's shared association with courage may explain their inclusion here, Elizabeth assigns much greater meaning than a single passion to both organs. Omitting the brain altogether, Elizabeth gives the stomach significance equal to that of the heart in defining selfhood and personal, gender, class, and national identity.[10]

The heart was thus not the only organ other than the brain that was linked to the mind and self. Complementing Stevens's exploration of the nexus of heart, brain, and mind with that of the belly, brain, and mind, I begin with the opening scene of Shakespeare's *Coriolanus* and the most famous Renaissance example of the Fable of the Belly, a variation on the metaphor of the body politic that was used to naturalize economic and political structures by comparing them to corporeal organs and systems. Shakespeare's Menenius Agrippa tells the Fable of the Belly in an attempt to calm the disgruntled citizens of Rome who believe the patricians guilty of hoarding grain in a time of dearth and making laws that worsen the plight of the poor. As Agrippa tells the tale, the body's members one day "[r]ebell'd against the belly," accusing it that "only like a gulf it did remain/ I' th' midst a' th' body, idle and unactive,/ Still cupboarding the viand, never bearing/ Like labor with the rest," while the "other instruments" who "mutually participate, did minister/ Unto the appetite and affection common/ Of the whole body" (1.1.97–105).[11] The response to this accusation is a physically detailed expansion of Shakespeare's sources:

> Your most grave belly was deliberate,
> Not rash like his accusers, and thus answered:
> "True is it, my incorporate friends," quoth he,
> "That I receive the general food at first
> Which you do live upon; and fit it is,
> Because I am the store-house and the shop
> Of the whole body. But, if you do remember,
> I send it through the rivers of your blood,
> Even to the court, the heart, to th' seat o' th' brain,
> And, through the cranks and offices of man,
> The strongest nerves and small inferior veins
> From me receive that natural competency

> Whereby they live. And though that all at once"—
> You, my good friends, this says the belly, mark me
> [. . .] "Though all at once cannot
> See what I do deliver out to each,
> Yet I can make my audit up, that all
> From me do back receive the flour of all,
> And leave me but the bran." (1.1.128–141, 143–146)

When asked to explain the moral of the fable, Agrippa asserts that the senators of Rome are the "good belly" while the citizens are "the mutinous members." He then adds: "disgest things rightly/ Touching the weal a' th' common, you shall find/ No public benefit which you receive/ But it proceeds or comes from them to you,/ And no way from yourselves" (1.1.148–154).

Agrippa clearly employs the fable as a piece of political propaganda, using the physiological importance of the belly to overall bodily health in order to defend the Roman patricians by metaphorically aligning them with the belly as distributer of nourishment rather than consumer of resources. The propagandistic function of this and other belly fables has received considerable scholarly attention, and Shakespeare's version has been interpreted as evidence of both democratic and antidemocratic sentiment.[12] I am particularly interested here, however, in what some of the details Shakespeare includes in his telling of the tale suggest about the psychological aspects of the belly and its relationship to the brain and brain function. That the belly is a psychosomatic site within the body and indicative of the mind's highly diffused embodiment in early modern thought is suggested in Agrippa's attribution of personality traits to the belly in his telling of the tale. Personification of bodily regions and of individual parts, a common feature of medical as well as literary texts, has the effect of intensifying the psychological aspect of these organs and bodily systems while at the same time pointing to the multifaceted nature of personal identity. In the Shakespearean version of the fable, the belly is described as replying to his accusers "tauntingly" and with "a kind of smile" (1.1.110, 107), a description which additionally enables, according to Arthur Riss, a stage joke during which the actor playing Agrippa may well have played with the folds of skin covering his abdomen to illustrate how the belly smiles.[13] The belly is also very specifically described as "grave" and "deliberate"; this description of course contributes to the ideological purpose of contrasting the patricians with the "rash" citizens, but the effect presents a belly that is especially thoughtful, seemingly with the capacity to think and reason on its own and even to undertake the complex intellectual work involved in making up an audit. Although the importance of all organs working together is a key element of the moral of the fable, the belly appears able to operate independently of the brain and heart and is presented here as the central decision maker of the body—just as the senate is in Rome—making judgments about the "natural competency" required by "the strongest nerves and small inferior veins." The heart and brain may be the "court" and "seat" (throne) of the body respectively, but they depend upon the centralized distributive function of the belly, just as Gershon suggests the brain in the head depends upon the second brain in the gut.

Agrippa importantly endows the personified belly in his tale with a capacity for speech, emphasizing this capacity through his use of direct rather than reported speech; early on he remarks to his audience, "look you, I may make the belly smile/ As well as speak" (1.1.109–110); and later he interrupts the belly's response specifically to add, "this says the belly, mark me." Alex Garganigo has remarked that Agrippa "mimics the belly's reply to the members with an implied bit of stage business many directors have interpreted as a burp or fart."[14] In addition to mimicking belly noises and thus adding humor (for the audience) and insult (for the citizens), Agrippa also, though, presents a belly that speaks in eloquent verse. By having Agrippa draw attention to his act of giving voice to the belly, Shakespeare may be alluding to the phenomenon of ventriloquism. This is, after all, a play very much about voices and about speaking through others. The tribunes, for example, are described as the "mouths" of the people (3.1.36), and to become consul, Coriolanus must ask the citizens to give him their "voices" (2.3.125–131).

Charles B. Davis writes that ventriloquism has, in recent years, "almost-become a dead metaphor; absorbed into academic jargon as a general term for any variety of speaking for or through a represented Other. The situations described by this metaphorical ventriloquism range from blatant cultural appropriation to the most well-intentioned attempts at giving a "voice" to marginalized groups."[15] He adds that the trope "has come to stand for the postmodern mistrust of both mimetic and sociopolitical representation."[16] Current theoretical use of the trope of ventriloquism, including Davis's, is based on the assumption that the ventriloquist speaks through a dummy or puppet, but from ancient times through the early modern period, ventriloquism was associated with the belly: the terms *ventriloquist*, *engastrimyth*, and *engastriloque* all literally mean someone who speaks out of or from the belly, stomach, or abdomen—a "belly-talker." Into the eighteenth century, "ventriloquism was deeply embedded in Christian discourses about demon possession, necromancy, and pagan idolatry," as the primary explanation for belly-talking was that the voice emanating from the belly came from a spirit that had possessed an individual.[17] Although increasing skepticism about the ability of supernatural forces to speak through individuals arose, discussions of ventriloquy remained torn, as Thomas Blount's entry under *ventriloquist* in his *Glossographia* (1656) illustrates: "one that has an evil spirit speaking in his belly, *or* one that by use and practice can speake as it were out of his belly, not moving his lips."[18]

Interpretations of ventriloquism reveal much about attitudes toward the body-mind and belly-brain relationship. Ventriloquism as belly-talking intensifies the concept of body language, powerfully highlighting the infusion of soma with psyche and stressing the psychological nature of the belly. In his depiction of Pantagruel's visit to the court of Master Belly (Sir Belly), Rabelais refers to an Italian woman "From whose Belly, we, as well as an infinite Number of others . . . have often heard the Voice of the Evil Spirit speak, low, feeble and small indeed; but yet very distinct, articulate and intelligible."[19] This evil spirit liked to be called *Curl à Pate* or *Cincinnatu'o* and would answer questions asked of him, though sometimes instead of answering he would let out "a rouzing Fart." Shakespeare likewise makes direct reference to the phenomenon of possession and belly-talking in *King Lear*, when Edgar, disguised as Tom O' Bedlam, describes how one of his devils, Hopdance, "cries" in his "belly" for "two white herring," and Tom responds to the devil's request by saying, "Croak not, black angel, I have no food for thee" (3.6.30–32). In this case, the cries supposedly made by the devil resemble the sounds made by an empty stomach. The belly is imagined in these examples as a source of speech separate from the brain and head, and that the spirit possessing the belly has a name and identity separate from the individual who is possessed exemplifies the possibility that the belly has a mind of its own.

Discussing possession more generally, David Hillman writes:

> Being possessed—having the (divine or human) other within you, means being taken over by something perceived as foreign, alien (for good or ill): being mad, transported, infatuated, in love. Having this alien entity within implies not being self-possessed, not being in one's right mind (and hence being *out* of one's mind), not being in control of oneself.[20]

Hillman adds to this comment in a footnote by citing Ruth Padel: "If you are *entheos* God 'inside,' you are *ekphrōn*. Your mind is 'out.'"[21] That demons specifically inhabit and speak from the belly in cases of ventriloquial possession implies, I suggest, that it is a mind in the belly that is displaced by the demon's inhabitation. The belief in a belly inhabited specifically by a demon combined with the strong connection between possession and madness provides an ancient example of the possibility raised by Gershon that the gut may have its own psychoneuroses and captures the misery experienced by individuals suffering from such diseases of the belly as those now known as Crohn's disease and irritable bowel syndrome. The concept of possession in the belly thus challenges later beliefs—pervasive even among neurogastroenterologists, Elizabeth Wilson claims—that the central nervous system is "the sole source of psychological influence" and that "[p]sychological events are confined to the higher neurological centers and transported down to the gut."[22] On the other hand, the alternate early modern interpretation of belly-talking as an act of vocal deception suggests a more centralized locus of consciousness in the head, and the interplay between the two interpretations reflects differing and changing opinions about the relationship among brain, mind, and belly and resembles later theorization of the relationship between the CNS and ENS.

Changing attitudes toward ventriloquism coincide with changing attitudes toward possession. In *Devil Theatre: Demonic Possession and Exorcism in English Renaissance Drama, 1558–1642*, Jan Frans van Dijkhuizen discusses differing interpretations of demon possession in the context of the religious tensions of the Reformation, and considers the possession controversy for insight into the nature of the body-mind relationship:

> Those writing in defence of the belief in possession and exorcism construed the self as porous, invadable by supernatural powers, and governed to a large extent by material, bodily processes. Writers who condemned possession and exorcism as an improbable fiction, by contrast, sought to sever the links between the physical and the spiritual, and attempted to formulate a disenchanted, unmagical view of human identity, which sees human beings as distinct from their surroundings, and from their bodies.[23]

Changing views about ventriloquism likewise relate to questions about the nature of the self and its relationship to the body and its environment. In his 1694 translation of Rabelais's description of Pantagruel's visit to the court of Master Gaster and his encounter with the Engastrimythes, who claim descent from the ancient Ventriloqui, "Southsayers, Enchanters, Cheats, who gull'd the Mob, and seem'd not to speak and give Answers from the Mouth; but from the Belly," Peter Motteux suggests, in his explanatory remarks, that these Engastrimythes are

> Parasites, and all those whom their hungry Bellies cause to say many things against their Consciences; so that they may be said to speak from the Belly: The word *Engastrymithe* also means one who by use and practice can speak as it were out of his Belly, not moving his Lips; and finally one who has an Evil Spirit speaking out of his Belly.[24]

Motteux here adds a figurative interpretation of belly-talking as speech motivated by hunger to interpretations of ventriloquism as evidence of deliberate trickery or demon possession, providing an example of the increasing metaphorization of bodily phenomena in the period traced by Hillman and Gail Kern Paster, among others, as well as setting the needs of the belly against conscience as opposing forces of body and mind. That the original etymological connection between ventriloquism and the belly has been lost also reflects the increasing emphasis placed on the head and the brain within it as the locus of the mind and its cognitive and affective processes and the corresponding diminishment of the belly's role in these processes.

In his theorization of the relationship between the brain and the second brain in the gut, Gershon writes:

> Evolution has played a trick. When our predecessors emerged from the primeval ooze and acquired a backbone, they also developed a brain in the head and a gut with a mind of its own. The organism could thus attend to more attractive things, like finding food, escaping destruction, and having sex with other organisms.[25]

Gershon likewise notes that "Descartes may have said, 'I think, therefore I am,' but he only said that because his gut let him. The brain in the bowel has got to work right or no one will have the luxury to think at all."[26] With the exception of the reference to evolution, Gershon's explanation of a key function of the second brain resembles Galen's explanation for the winding structure of the intestine. In many of the "voracious animals," Galen notes, the "intestine is straight" and of uniform width, with the result that "these animals both feed continually and as incessantly eliminate." They thus lead "a life truly inimical to philosophy and music, as Plato has said, whereas nobler and more perfect animals neither eat nor eliminate continually," because of the winding coils of the intestine.[27] Early modern medical writers perpetuated this interpretation of the importance of the intestine to human activity. As Helkiah Crooke expressed it in his 1615 *Microcosmographia*, the "guttes" are convoluted to prevent "insatiable gulosity or rauenousnes, which would as *Plato* saith, interrupt all good and liberall learning."[28] Ambroise Paré similarly explains that the

> length of the guts, is seven times more than the length of the whole body; to this length they have windings, least the nourishment should quickly slide away, and least men should bee withdrawne by gluttony from action and contemplation. For so wee see it comes to passe in most beasts, which have one Gut stretched straight out from the stomacke to the fundament; as in the Lynx and such other beasts of insatiable gluttony, alwayes, like plants, regarding their food.[29]

The guts were thus ascribed a fundamental role (so to speak) in cognitive processes, including philosophizing, and artistic creativity—including that exemplified in music and, one could add, Shakespearean drama—that we associate with the mind and with being human.

Not only did the structure of the guts allow for contemplation and action, but the digestive process was understood to contribute directly to the production of the material underpinnings of thought as well as emotion. In Agrippa`s fable, the belly is accused of not "minister[ing]/ Unto the appetite and affection common/ Of the whole body," and its defense thus implies that its role in the distribution of aliment is crucial to both the "appetite" (desire for food or "bent of mind toward the attainment of an object or purpose") and "affection" (mental state/ emotion/feeling) of the body as a whole.[30] The most obvious example of what has been described as the early modern materialist psychology or psychophysiology inherited from Galen is the role digestion was believed to play in the production of humors. According to Galenic humoral theory, digestion took place primarily through heat (rather than chemical reaction), and, prior to general acceptance of William Harvey's theory of blood circulation (first published in 1628), it was believed to produce blood on a daily basis. In the stomach, ingested food was converted into chyle, which was then conveyed by means of the small intestine and meseraic veins to the liver, where it was made into blood for assimilation by the rest of the body, while excrements arising from the digestive process were removed by the large intestine and other excretory organs. The humors produced during digestion—yellow bile, black bile, phlegm, and blood—were classified as hot or cold and wet or dry. The digestive process was linked to mental and emotional experience since the proportion of these humors within an individual affected temperament, making a person choleric, melancholic, phlegmatic, or sanguine, respectively. As Gail Kern Paster stresses, the "forces of cold, hot, wet, and dry constituted the material basis of any living creature's characteristic appraisals of and responses to its immediate environment: they altered the character of a body's substances and, by doing so, organized its ability to act or even to think."[31] Foods, like the humors they gave rise to, were categorized as hot or cold and moist or dry. For example, Paris claims in *Troilus and Cressida* that doves had an aphrodisiacal effect since eating them "breeds hot blood, and hot blood begets hot thoughts, and hot thoughts beget hot deeds, and hot deeds is love" (3.1.128–130). Humoral balance, achieved through purgation and diet, was considered fundamental to overall physical and mental health.[32]

Digestion was also a crucial stage in the production of spirits, thin substances resulting from various stages of refinement: the vital spirits were made in the heart from the blood produced in the liver, while the animal spirits were made in the brain from vital spirits. Spirit, as Robert Burton notes in *The Anatomy of Melancholy*, served as "the instrument of the soul, to perform all his actions; a common tie or medium between the body and the soul, as some will have it; or as Paracelsus, a fourth soul of itself."[33] Spirits were directly involved in emotional experience and in acts of memory related to that experience, and the quality of blood produced during digestion profoundly influenced these experiences. Not only did "hot blood" give rise to "hot thoughts," as noted by Paris, but, as Lemnius explains, thick and "grosse" blood produced in the digestive process results in "corpulent and stronge spirits," and those with such spirits will consequently "beare a grudge in memory a long tyme, and not easely forgette those motions and heddines that they once take"; contrastingly, those with thin blood and "slender spirits" that "soone passe away" become enraged quickly but do not remain so for long.[34] As well, impaired digestive function could result in the production of "unpure" spirits and harmful vapors that might "disturbe and hurt the brayne & mynde."[35]

The ancient perception of a direct and reciprocal connection between the belly and mental processes is perpetuated medically in the early modern period through the continued predominance of Galenic theory and linguistically through translation of proverbial expressions. Juan Huarte, for example, provides (in a text itself translated into English) a saying alleged by Galen that "*A grosse bellie makes a grosse understanding.*" Huarte explains:

> this proceeds from nothing else, than that the brain and the stomacke are vnited and chained together with certaine sinewes, by way of which they interchangeably communicat their dammages. And contrariwise, when the stomacke is drie and shrunke, it affoords great aid to the wit, as we see in the hungerstarued, and such as are driuen to their shifts, on which doctrine (it may be) *Persius* founded himself, when he said, That the belly is that which quickens vp the wit.[36]

In Huarte's interpretation of Persius's doctrine relating the condition of the belly to brain function, a hungry belly sharpens the wit, not least to seek remedies for a potentially fatal shrinking belly, while a gluttonous one impairs mental faculties. In Rabelais's description of Pantagruel's visit to the court of Master Gaster, he critically observes that the whole world is busy serving the belly, described as the "great Master of Ingenuity" and "the first Master of Arts in this World," who in return for this service invents "All Arts, Machines, Trades, Engines, and Crafts."[37] In Rabelais's version, the belly is not just the stimulus of mental activity but appears to be its source.

The belly was often used metonymically to refer to the body in its capacity for food and indulgence, and *belly* could simply mean "a glutton."[38] Similarly, *belly-god* was a term used to refer to someone who made a god of his belly, following belly-doctrine rather than church doctrine and attending to corporeal appetite rather than the injunctions of reason. This negative interpretation of the belly is highlighted in *Coriolanus* by the first citizen, who refers to the "cormorant belly," the "sink" of the body in contrast to the much more positive "kingly-crowned head, the vigilant eye,/ The counsellor heart, the arm our soldier,/ Our steed the leg, the tongue our trumpeter" (1.1.120–121, 115–117). Many early modern writers warned of the dangers posed by gluttony to the brain and understanding by positioning the belly in opposition to the brain. Edward Reynolds, for example, strongly advises against "intemperate excesse," against becoming "in the Apostle's phrase . . . *lovers of pleasure rather than lovers of God*," making "our belly the grave of our Soule, and the dungeon of our Reason, and let[ting] our *Intestina* as well morally as naturally farre exceed the length of the whole Man besides."[39] Chastizing gluttonous meat eaters, Thomas Tryon writes: "though many of our *English Belly-Gods* suppose Flesh to be most mighty in its operation; but they are grosly mistaken; and no wonder, since their *Brains*, *Wits*, and *Understandings*, are, through Intemperance, Superfluity, and Gluttony, *sunk into their Bellies*."[40]

These articulations of the antagonistic relationship between the flesh (belly) and reason (brain) highlight the negative effect of gluttony on the brain, but do so in a way that points to brain-like function in the belly through a perceived exchangeability (though certainly not a desired one) between the brain and gut. This exchangeability may be one reason why Thersites claims in *Troilus and Cressida* that the brain and guts have traded places in Ajax's body. Speaking scornfully to Achilles of Ajax's stupidity, Thersites says: "I will buy nine sparrows for a penny, and his *pia mater* is not worth the ninth part of a sparrow. This lord. . . . who wears his wit in his belly and his guts in his head" (2.1.70–73). A pun on "Ajax" as a jakes, or privy, is another, related reason for the comparison. Although Thersites's comment is clearly meant as an insult, this and the other comments, by imagining the brain, with its reasoning faculty, in the belly, contribute to an early modern version of the "second brain" concept of neurogastroenterology.

This imagined exchangeability of brain and guts additionally underscores structural and functional similarities between them, similarities relevant to understandings of the mind. In particular, the winding structure of the intestinal tract means that more than any other organ it most closely resembles the brain. As Crooke explains in *Microcosmographia*, the ancients and those after them have, according to Vesalius, "excellently compared" the "diuers orbicular circumuolutions and circular ruts" of the brain "to the gired windings of the guts when the kell [omentum] is taken off."[41] Furthermore, while some claim that these convolutions have no particular use, "we with *Galen* do determine that their vse is to secure the vessels, that going together with the *Pia mater* they might carry vnto the braine nourishment and life, and not be in danger of breaking in the perpetuall motion and agitation of the brain."[42] A similar opinion held that the venal and arterial blood arriving in the brain needed to remain in the winding vessels to allow time for further elaboration and time to "purge it from those excrements which are gathered and heaped vp in these Conuolutions" so that the brain could use it for "his nourshment and also for the procreation of Animall spirits."[43] As indicated here, the brain not only resembles the guts in its convolutional structure but also in the elaborative, nourishing, and excretory function of those convolutions. In this context, it is worth noting the medical use of the term *buttocks* by Crooke to describe two parts of the brain, called "small buttocks" by Galen because "they carry the representation of two buttocks ioyned together." These buttocks serve to support the weight of the brain, and when the "After-braine is remooued, the passage out of the third ventricle into the fourth conspicuously appeareth which passage resembleth the very fundament between the buttocks." [44] While having one's brains in one's guts could imply stupidity, so could not having guts in one's brains:

in Peter Motteux's translation of the prologue of *The Fifth Book of the Works of Francis Rabelais, M.D.*, a "sot" is defined as "a Dunce, Dullard, Jolthead, Gull, Wittal, or Noddy, one without Guts in his Brains, whose Cockloft is unfurnish'd, and in short, a Fool."[45]

John Sutton has examined the significance of frequent early modern descriptions of the brain as "spongy," noting that "like brains, sponges were peculiar because of their porous nature, able both to absorb and erase" and often used "to wipe the erasable leaves of table-books, which became increasingly common in England from the 1580s onwards."[46] An example not discussed by Sutton comes from Crooke, in his explanation of how the pituitary glandule "receiueth the excrements in manner of a sponge as they fall from the braine, which excrement it not onely emptieth into the palate, but also some fals downe by his sides through those holes which are bored in the Basis of the Scul."[47] The brain's sponginess is also highlighted in the description of the body in Sylvester's translation of Du Bartas's *Divine Weeks and Works*:

> The Nose, no lesse for use then beautie makes;
> For, as a Conduit, it both gives and takes
> Our living breath: it's as a Pipe put up,
> Whereby the moist Braines spongie bone doth sup
> Sweet smelling fumes.[48]

In a poetic comparison of brain and digestive organs, the brain's "spongie bone" is described as specifically supping on air as a means of internalizing the external in the form of breath and the sense of smell. This connection is even further consolidated in the poem when the brain is described as having:

> That curious Maze, that admirable Nett,
> Through whose fine folds the spirit doth rise and fall,
> Making its powers, of *Vital*, *Animal*:
> Even as the Blood and Spirits, wandering
> Through the *preparing vessels* crooked Ring,
> Are in their winding course concoct and wrought,
> And by degrees to fruitfull *Seed* are brought.[49]

As in the example from Crooke, the digestive aspect of brain function is stressed, suggesting that the idea of the second brain in the gut has a correlate in the idea of a second gut in the brain, and indeed digestion was a process imagined to occur throughout the body as various parts engaged in what was frequently described as the third concoction during which the parts assimilated what nourishment they required.

Importantly, the adjective *spongy* was also applied to the bowels, further highlighting the similarity of function shared by them and the brain.

Again in *Troilus and Cressida*, Hector gives his opinion that the Trojans should return Helen to the Greeks and end the war, remarking: "Though no man lesser fears the Greeks than I/ As far as toucheth my particular,/ Yet, dread Priam,/ There is no lady of more softer bowels,/ More spungy to suck in the sense of fear,/ More ready to cry out, 'Who knows what follows?'/ Than Hector is" (2.2.8–14). Hector's comment emphasizes the role played by the bowels in absorbing what is external, whether nutriment, emotions like fear, or ideology like that perpetuated by Hector's gendering of both the bowels and fear. These connections between the brain and belly and between brain function and digestion are further highlighted in early modern expressions of how knowledge is acquired and processed. After telling the Fable of the Belly, for example, Agrippa claims that if the citizens of Rome "disgest things rightly" they will realize the many benefits bestowed upon them by the patricians, underscoring the powerful connection perceived to exist between the belly and learning. Discourse was repeatedly imagined as something assimilated (to varying degrees) into the mind by way of the digestive tract. In *Julius Caesar*, Cassius says of Casca that his "rudeness is a sauce to his good wit,/ Which gives men stomach to disgest his words/ With better appetite" (1.3.300–302), while of written texts, Francis Bacon famously advises: "Some bookes are to bee tasted, others to bee swallowed, and some few to bee chewed and disgested: That is, some bookes are to be read only in partes; others to be read, but cursorily, and some few to be read wholly and with diligence and attention."[50]

Spenser's depiction of Error in Book One of *The Faerie Queene* literalizes digestion of language and text by focusing on its reversal through vomiting. Rather than digesting texts, Error "spew[s]" a "floud of poison horrible and blacke," and her "vomit full of books and papers was" (1.1.20).[51] While generally overlooked in discussion of this commonplace trope of books as food and reading or learning as eating and digesting, there were actual instances of literal ingestion of the written word. For example, in the 1573 siege of Sancerre, people were forced by hunger to eat leather and, when that was gone, they

> turned to trying parchment "not only white blank parchment, but also letters, title-deeds, books printed and hand written, having no difficulty in eating even those a hundred or a hundred and twenty years old." They were soaked, chopped, boiled for a day and a half, and then they were fricasseed like tripe, or cooked with herbs and spices.[52]

At the other culinary extreme, scholars studying cookbooks and banqueting practices of the early modern period have described how some banqueting dishes included poems impressed into sugar or marzipan. Even when no text was included, these dishes, also called "conceits," took the shape of other objects and often contained political messages, which could perhaps be similar to the Fable of the Belly, the moral of which Agrippa wishes the hungry citizens to "disgest" correctly.[53]

These examples of physical and metaphorical ingestion of texts foreground the physiological and psychological role of the digestive tract in the relationship of the self to the world. In *Shakespeare's Brain*, Mary Thomas Crane draws upon a branch of cognitive theory to argue that the "power of culture to shape individual selves must be filtered through the material, biological constructs of the brain," the physical shape of which "social and cultural interactions" may materially alter.[54] In the above examples and numerous others, however, words and their cultural values are substances one ingests, digests, and assimilates, or they might be expelled from the stomach or remain undigested, and they can have a directly visceral impact on the body's digestive system as do social experiences. Expressions like "to pierce one's stomach," "to sink (deep) into one's stomach," or "to stick in one's stomach," which according to the *Oxford English Dictionary* were used of things that made a lasting impression on the mind, especially a painful one, suggest that the belly plays more than a metaphorical role in the self's interaction with culture and environment; the stomach in these expressions is perceived as a locus of mind and as susceptible to psychological trauma.[55] Using recent research on the brain in the gut, Elizabeth Wilson makes a similar observation about the digestive tract's importance to the relationship between the self and society: "What the outside world engenders in the psychological sphere is relations to others, and through this the development of the self . . . . These relations to others are psychologically generative only to the extent that they are internalized (ingested, absorbed, excreted)."[56] She adds the suggestion that maybe "ingestion and digestion aren't just metaphors for internalization; perhaps they are 'actual' mechanisms for relating to others." In Wilson's opinion, scientific research appears to support this view: "The large stocks of serotonin in the gut, the morphological similarities between gut neurons and brain neurons, and the clearly psychological character of gut function all suggest that it is not just ideation that is disrupted in depression; it is also the gut."[57]

Early modern perspectives on the body-mind relationship and on the belly-brain relationship ought not, then, be considered important only as the transition from what Gershon describes as "Medieval ignorance" to Enlightenment views on the gut. Renaissance medical and cultural belief about the belly's contribution to mind, while not informed by modern research on such things as serotonin and neurotransmitters, but by medieval and ancient beliefs, nonetheless suggest awareness of a relationship between the belly and brain resembling that proposed by researchers who consider the gut to have brainlike capacities of its own. Gershon's suggestion that "the brain in the bowel may also have its own psychoneuroses" and that this "new concept" may "turn out to be as revolutionary and as hopeful as Copernicus's discoveries" again misses the point that the medical and scientific revolution of which Copernicus was a part (as was Descartes) contributed to the diminished role assigned to the belly in cognitive and emotional experience prior to the "discoveries" of recent research into the gut and the body-mind relationship.[58]

NOTES

I would like to thank Laurie Johnson, John Sutton, Evelyn Tribble, and the participants of The International Shakespeare Association Congress "Body-Mind" Seminar (2011) for their helpful comments on earlier versions of this paper.

1. Michael D. Gershon, *The Second Brain: A Groundbreaking New Understanding of Nervous Disorders of the Stomach and Intestine* (New York: Harper, 1998), xiii, 17, xiv. Elizabeth A. Wilson notes, in *Psychosomatic: Feminism and the Neurological Body* (Durham, North Carolina: Duke University Press, 2004), that the ENS "has been variously named 'the brain of the gut,' 'the enteric minibrain,' and 'the second brain,'" but cautions that assertions of the independence of the ENS from the CNS tend to overstate the case; she instead describes the relationship as "largely independent yet indirectly connected" (34, 36).
2. Wilson, 43, 33.
3. Gershon, xv.
4. Scott Manning Stevens, "Sacred Heart and Secular Brain," in *The Body in Parts: Fantasies of Corporeality in Early Modern Europe*, ed. David Hillman and Carla Mazzio (New York: Routledge, 1997): 263–282; esp. 267.
5. Stevens, 271.
6. Stevens, 267.
7. Richard Broxton Onians, *The Origins of European Thought about the Body, the Mind, the Soul, the World, Time, and Fate* (Cambridge: Cambridge University Press, 1951), 85, 89.
8. Levinus Lemnius, *The Touchstone of Complexions*, trans. Thomas Newton (London, 1581; Early English Books Online), fol. 23v.
9. Carole Levin, *"The Heart and Stomach of a King": Elizabeth I and the Politics of Sex and Power* (Philadelphia: University of Pennsylvania Press, 1994), 144.
10. For more on Elizabeth's use of *stomach* in these contexts, see Jan Purnis, "The Stomach in Early Modern Emotion," *University of Toronto Quarterly* 79.2 (2010): 800–818.
11. All Shakespeare quotations are from *The Riverside Shakespeare*, ed. G. Blakemore Evans (Boston: Houghton Mifflin, 1974): hereafter cited in text.
12. Leonard Barkan, for example, notes that in the Fable of the Belly tradition, the belly "always stands for some seemingly useless element of society, and the moral is not only that the nobles, or senators, or clergy, are indispensable but also that the amputation of any one part of the body politic is fatal to all the others." See Leonard Barkan, *Nature's Work of Art: The Human Body as Image of the World* (New Haven, Connecticut: Yale University Press, 1975), 96. Michael Schoenfeldt also observes: "Fantasizing that a society's resources naturally trickle down from its most to its least privileged members, Agrippa uses the physiological centrality of the stomach to mystify a doctrine of social inequality and to obscure the actual labor that is part of the production and distribution of provisions." See Michael

Schoenfeldt, "Fables of the Belly in Early Modern England," in *The Body in Parts: Fantasies of Corporeality in Early Modern Europe*, ed. David Hillman and Carla Mazzio (New York: Routledge, 1997): 243–262; esp. 248–249. For more on the Fable of the Belly tradition generally, see David Hale, *The Body Politic: A Political Metaphor in Renaissance English Literature* (The Hague: Mouton, 1971), and Annabel Patterson's chapter "Body Fables" in her *Fables of Power: Aesopian Writing and Political History* (Durham, North Carolina: Duke University Press, 1991): 111–137. For a brief overview of contrasting readings of *Coriolanus* and an insightful analysis of Shakespeare's engagement with representational politics in the play, see Oliver Arnold, *The Third Citizen: Shakespeare's Theater and the Early Modern House of Commons* (Baltimore: Johns Hopkins University Press, 2007), 192, 197–214.

13. Arthur Riss, "The Belly Politic: *Coriolanus* and the Revolt of Language," *ELH* 59 (1992): 53–75; esp. 61.
14. Alex Garganigo, "*Coriolanus*, the Union Controversy, and Access to the Royal Person," *Studies in English Literature 1500–1900* 42.2 (2002): 335–359; esp. 351. Garganigo reads this possible fart in the context of contemporary politics. See also Maurice Hunt, "The Backward Voice of Coriolanus," *Shakespeare Studies* 32 (2004): 220–239; esp. 222, and A. Crunelle, "*Coriolanus*: The Smiling Belly and the Parliament Fart," *ANQ: A Quarterly Journal of Short Articles, Notes, and Reviews*, 22.1 (2009): 11–16. In Riss's opinion, Agrippa "belches to illustrate how the belly can speak" (61).
15. C. B. Davis, "Reading the Ventriloquist's Lips: The Performance Genre Behind the Metaphor," *The Drama Review* 42.4 (1988): 133–156; esp. 133.
16. Davis suggests that the "semiotic complexity of ventriloquist performances makes it an apt analogy for the precarious navigation of multiple subject positions faced by each speaker, particularly in pluralistic or intercultural contexts" (151).
17. Leigh Eric Schmidt, "From Demon Possession to Magic Show: Ventriloquism, Religion, and the Enlightenment," *Church History* 67.2 (1998): 274–304; esp. 277.
18. Schmidt, 280–281.
19. François Rabelais, *Pantagruel's Voyage to the Oracle of the Bottle. Being the Fourth and Fifth Books of the Works of Francis Rabelais, M. D*, trans. Peter Motteux (London, 1694; Early English Books Online), 4.58.228.
20. David Hillman, *Shakespeare's Entrails: Belief, Scepticism, and the Interior of the Body* (Basingstoke: Palgrave Macmillan, 2007), 120–121.
21. Hillman, 215 n.6.
22. Wilson, 40.
23. Jan Frans van Dijkhuizen, *Devil Theatre: Demonic Possession and Exorcism in English Renaissance Drama, 1558–1642* (Cambridge: D. S. Brewer, 2007), 26. Dijkhuizen suggests that the Church of England, in contrast to the Catholics (and also Puritans), wished "to downplay the theological role of the body" (29).
24. Rabelais, 4.58.227, lix.
25. Gershon, xiii.
26. Gershon, xv.
27. Galen, *On the Usefulness of the Parts of the Body*, trans. Margaret Tallmadge May (Ithaca, New York: Cornell University Press, 1968), I: 240.
28. Helkiah Crooke, *Microcosmographia: A Description of the Body of Man* (London, 1615; Early English Books Online), 103.
29. Ambroise Paré, *The Collected Works of Ambroise Paré*, trans. Thomas Johnson (New York: Milford House, 1968), 108.

30. *OED*, n. "appetite" (1a); n.1. "affection" (1a).
31. Gail Kern Paster, *Humoring the Body: Emotions and the Shakespearean Stage* (Chicago: University of Chicago Press, 2004), 13.
32. In addition to Paster's *Humoring the Body*, for more on humoral theory, see also Michael Schoenfeldt, *Bodies and Selves in Early Modern England: Physiology and Inwardness in Spenser, Shakespeare, Herbert, and Milton* (Cambridge: Cambridge University Press, 1999) and Mary Floyd Wilson, *English Ethnicity and Race in Early Modern Drama* (Cambridge: Cambridge University Press, 2004).
33. Robert Burton, *The Anatomy of Melancholy*, ed. Holbrook Jackson, intro. William H. Gass (New York: New York Review Books, 2001), 1.1.148.
34. Lemnius, 14v.
35. Lemnius, 10v.
36. Juan Huarte, *The Examination of Mens Wits*, trans. Richard Carew (London, 1594; Early English Books Online), 26.
37. Rabelais, 4.58.227, 4.57.222, 4.57.225.
38. *OED*, n. "belly."
39. Edward Reynolds, *A Treatise of the Passions and Faculties of the Soule of Man* (London, 1640; Early English Books Online), 164.
40. Thomas Tryon, *The Way to Health, Long Life and Happiness: Or, A Discourse of Temperance* (London, 1691; Early English Books Online), 291.
41. Crooke, 456.
42. Crooke, 457.
43. Crooke, 457.
44. Crooke, 468. For an example of Crooke's use of "buttocks," see 467, FG in fig. 10.
45. François Rabelais. *The Fifth Book of the Works of Francis Rabelais, M.D. Containing the Heroic Deeds and Sayings of the Good Pantagruel*, trans. Peter Motteux (London, 1694; Early English Books Online), A2.The *OED* offers this example for the connotation of *guts* as "substantial contents" (1c), but there is, I think, more going on here.
46. John Sutton, "Spongy Brains and Material Memories," in *Environment and Embodiment in Early Modern England*, ed. Mary Floyd-Wilson and Garrett A. Sullivan Jr. (Houndmills, Basingstoke, Hampshire: Palgrave, 2007): 14–34; esp. 17.
47. Crooke, 470.
48. Du Bartas, Guillaume de Saluste, *Divine Weeks and Works of Guillaume de Saluste, Sieur de Bartas*, trans. Joshua Sylvester, ed. and intro. Susan Snyder (Oxford: Clarendon; New York: Oxford University Press, 1979), 1: 277–278, lines 559–563.
49. Du Bartas, lines 678–684.
50. Francis Bacon, "Of Studies" (1597), in *The Broadview Anthology of Seventeenth-Century Verse & Prose*, ed. Alan Rudrum, Joseph Black, and Holly Faith Nelson (Peterborough, Ontario: Broadview, 2004): 45–46.
51. Edmund Spenser, *The Faerie Queene*, ed. Thomas P. Roche, Jr., with assistance of C. Patrick O'Donnell (London: Penguin, 1987).
52. Andrew Cunningham and Ole Peter Grell, *The Four Horsemen of the Apocalypse: Religion, War, Famine and Death in Reformation Europe* (Cambridge: Cambridge University Press, 2000), 229.
53. See for example, Sidney Mintz, *Sweetness and Power: The Place of Sugar in Modern History* (New York: Penguin, 1985): 87–95; or Kim F. Hall, "Culinary Spaces, Colonial Spaces: The Gendering of Sugar in the Seventeenth Century," in *Feminist Readings of Early Modern Culture: Emerging Subjects*, ed. Valerie Traub, M. Lindsay Kaplan, and Dympna

Callaghan (Cambridge: Cambridge University Press, 1996): 168–190; esp. 172–178.

54. Mary Thomas Crane, *Shakespeare's Brain: Reading with Cognitive Theory* (Princeton: Princeton University Press, 2001), 23, 10.
55. *OED*, n. "stomach," (6c.)
56. Wilson, 44.
57. Wilson, 45.
58. Gershon, xiv.

Afterword

David Hillman and Carla Mazzio

This volume has drawn together approaches to Shakespeare and early modern drama that engage, in historically specific ways, with questions about embodied cognition newly invigorated by developments in cognitive psychology and neuroscience, science studies, and philosophy. As Johnson, Sutton, and Tribble point out in their helpful introduction, such a project requires us to tread carefully in the domain of the "body-mind problem": "we suggest that the labels 'Cartesian,' 'pre-Cartesian,' and 'post-Cartesian' can mislead as well as inform our views of the body-mind." Thus, rather than serving as an explicit response to dualist systems or binary models operative in early modernity or contemporary criticism, the volume takes the already dynamic "body-mind"—and the rich and variegated set of issues involved—as its point of departure. As such, the essays evince an increasing awareness among literary and cultural historians that early modern writers not only treated the mental world as fully interactive with the corporeal—witness Montaigne's "close stitching of mind to body, each communicating its fortunes to the other"—but also understood both the body and the mind as fully interactive with other bodies and minds, with "embodied objects," and with a myriad of cosmic and environmental factors as well as unknowable (e.g., supernatural) beings. Personhood, in this world, is an inclusive notion.

Indeed, what becomes abundantly clear in reading these essays is that when we speak of "body" (and similar things can be said of "mind" or "body-mind"), we need to start thinking of the terms as referring not merely to the entity, bounded by the skin, stretching from the tips of the toes to the crown of the head, but of a far more porous and inclusive entity that is constantly interacting with everything around it. And the issue works both ways: that is to say, it is not just that "we" include the world around us, but that, as Gail Kern Paster puts it, "the properties of the world are never innocent of human intention or consciousness." So that what we are looking at here might better be termed the mind-body-world problem.

As Lenin is said to have remarked, everything is connected to everything else. But if we believe that the mind is so embedded or anchored in the world, if it is not just the mind that does the thinking, there are profound implications for the way we conceive of subjectivity and agency—and thus of responsibility and ethics. It is these implications that early modern drama displays and interrogates, and, broadly speaking, that these essays take on. For "Mind," in these essays, is an inclusive concept, incorporating body (touch and the other senses, the "intellectual gut" with its gastrointestinal neurotransmitters, extension, action), surroundings, objects. As a result, as Laurie Johnson puts it, "the unit of analysis is [now] the context itself." Modern science is growing ever more aware of the ways in which abstract mathematical, philosophical, and psychological concepts, indeed language itself—the basic tool of mentation—all are intimately connected with (as well as based on and continually benefiting from) corporeal experience. As Emma Firestone summarizes, "much evidence now exists to support this intuition" (the flow, we should note, is implicitly from the pre-existing intuition to the empirical evidence). There is no such thing as pure mentation (or pure bodiliness).

Perhaps that is one reason why so many of these essays are repeatedly interested in plays where adultery is at issue (*The Winter's Tale*, *Othello*, *A Woman Killed with Kindness*, *Hamlet*, *Much Ado About Nothing*): the mind is always at risk of being adulterated by the flesh (and perhaps the flesh of being "sullied" with the mind), as if what is being looked at is a kind of infidelity (between male and female, inner and outer, matter and spirit). As Darryl Chalk writes, "Leontes's suspicion is bound up in this play and in early modern treatises with the medical understanding of love and jealousy as kinds of infection, diseases of the body-mind received through the eye in an instant by contagion." And indeed permeability, infection, contagion are a common theme of many of these essays. Yet, paradoxically, it may also be the case that, as Michael Schoenfeldt and others suggest, early modernity is the period to which we can best ascribe the "modern" paradigm of the body as an inviolable and relatively fixed container, the shift away from environmental personhood.

This raises two questions that emerged from a reading of this volume: how or why does early modern drama, and Shakespearean drama in particular, function as a particularly generative site for thinking and feeling our way through issues surrounding the "body-mind"? And when and why is it that the "body-mind" emerges, in particular moments of history and criticism, as a problem? To touch on the latter question first, it is worth noting that the body-mind debate, of course, has never been settled; it is just that the problem becomes particularly intense at certain historical junctures—when, in particular, scientific findings make it hard not to be acutely curious about these issues. It is when the pressure to explain these findings is ratcheted up that it becomes a *problem*. The late sixteenth and early seventeenth century, with Baconian and Cartesian epistemologies hovering in the wings, appears to have been one such period; our own time of rapid technological and biomedical advance (including new ways of visiting the living interior of the body which we could parallel with the advances in anatomical discoveries in the sixteenth century) has placed renewed pressure on these questions. In both cases, literary and artistic endeavors find something provocative and useful, even thrilling, in the scientific process as well as in the findings. But what we need to bear in mind (or should we say in body-mind) is that this engagement always took place under the sign of curiosity and resistance, not of cooptation, and that—especially in a world so dominated by a scientific episteme as ours is—literary critics would do well to follow the example of these early modern writers who do not simply accept the paradigms of empirical science as given. These essays remind us that while it is surely important and can be illuminating to set early modern literature in the context of early modern or modern medical/physiological/natural-historical paradigms, it is also important to shy away from using these paradigms as keys to unlock the meanings of, for example, a Shakespeare play. For such a procedure bears the risk of flattening out or stripping some of the complexity and luminosity of the texts, as well as of missing one of the lessons we can learn from reading early modern literature—an epistemological lesson about the skepticism with which we should treat all forms of so-called knowledge, including empirical knowledge. Literature—and Shakespeare more than most—never passively accepts the explanations it encounters; nor should literary criticism.

Early modern drama, which at once necessitates and endlessly comments upon conditions of embodiment as a part and parcel of networks of meaning, thinking and feeling, is of course a particularly rich arena for exploring relationships among bodies, minds, and worlds through which *dramas* of knowledge come into view. The so-called "humours" comedy goes to one extreme in representing embodied cognition, making its characters seem near-helpless victims of our humoral "complexions." Shakespeare, among other early dramatists, takes embodied cognition to a different—a much more unsettling—place, making the mind-body problem truly a *problem*, one that cannot ever be resolved but that, on the contrary, generates profound questions of subjectivity and agency. Shakespearean drama does not, of course, hold a mirror up to neurons (or the early modern equivalents thereof), or the technical process through which embodied cognition "works." But it stages whole worlds in which problems and conundrums about embodied cognition inform dramas of human interaction. Particularly gripping is the notion of Shakespeare's theatre as perhaps the ultimate example *of* distributed cognition or an extended system or web; a place where we can see the ways in which play, actors, audience, and wider culture are all interlinked in a quasi-haptic environment: the drama *touches* us in every sense of the word. The idea of cognition distributed across time as well as across bodies and worlds offers a possible way to begin to think—without appeals to universalism—about the enduring power of Shakespeare in our lives and in our body-minds.

At the same time, it is interesting to note how often the rejection of a divorce of the mind and body, in the Introduction and elsewhere in this volume, goes hand-in-hand with the insistence on re-marrying Shakespeare and his early modern milieu, as if the "genius" of the dramatist, the grandeur of his "spirit," rising above the "very age and body of the time" were analogous in some way to the old hierarchy of mind over matter. By organizing an approach to Shakespeare through the rubric of embodied cognition, what emerges is of course the importance of models of embedded personhood over the comparatively antiquated, no longer tenable notion of the solitary, monadic, single individual. Given that this is the case, it is worth reflecting upon the fact that—as students and scholars of early modern literature and culture—our minds are not only embodied but also deeply embedded; in this case, embedded within networks of socio-economic and institutional value that, on the one hand, still privilege Shakespeare as an icon of literariness, clearly separate from his fellow playwrights, even as the value of literary study hangs in the balance, and on the other, privileges science over that of the humanities. This is simply to highlight the stakes of the volume and to ask ourselves how our own scholarship is unwittingly as well as wittingly embodied and embedded at the present moment, even or especially in the process of working to make Shakespeare matter.

Contributors

Hardin Aasand is Professor of English and Chair of the Department of English and Linguistics at Indiana-Purdue University, Fort Wayne, Indiana. He is a member of the editorial team for the New Variorum *Hamlet* for the MLA. He edited *Stage Directions in Hamlet* (2003) and *The Winter's Tale* for the Internet Shakespeare Editions and Broadview Press (2014). He has published essays in *Studies in English Literature*, *Hamlet Studies*, and chapters in a number of volumes. He also has written review essays for *Renaissance Quarterly*, *Seventeenth-Century News*, *Medievalia et Humanistica*, and the *Sixteenth-Century Journal*.

Darryl Chalk is Senior Lecturer in Theatre at the University of Southern Queensland. He is coeditor with Laurie Johnson of *Rapt in Secret Studies: Emerging Shakespeares* (Cambridge Scholars, 2010). Publications of his work on contagion in Shakespearean drama and early modern culture include an article on *Timon of Athens* for *Early Modern Literary Studies* (Special Issue 19, 2009) and book chapters on *Troilus and Cressida* in *"This Earthly Stage": World and Stage in Late Medieval and Early Modern England,* Brett D. Hirsch and Christopher Wortham, eds. (Turnhout, Belgium: Brepols, 2010) and *Coriolanus* in the forthcoming Prague ISA World Congress collection. He is currently writing a monograph on this topic.

Emma Firestone (Ph.D. candidate) holds Bachelors of Arts from Harvard University (Folklore and Mythology) and the University of Cambridge (English) and an M.Phil from Cambridge (Medieval/Renaissance Literature). For Cambridge's Faculty of English, she is currently completing her doctoral thesis on the cognitive processes underlying audiences' understanding of and assent to a range of phenomena in Shakespeare's work. Emma has produced and directed several Shakespeare plays over the years and is trained in mime and *commedia dell'arte.* As well as supervising Cambridge undergraduates, Emma has taught Japanese students through Pembroke College's International Programmes Department since 2007. She lives in Denmark.

Mary Floyd-Wilson is Professor of English and Comparative Literature at the University of North Carolina at Chapel Hill. She is the author of *English Ethnicity and Race in Early Modern Drama* (2003) and *Occult Knowledge, Science, and Gender on the Shakespearean Stage* (2013). She has also coedited *Reading the Early Modern Passions: A Cultural History of Emotion* (2004) and *Environment and Embodiment in Early Modern England* (2007)

Lianne Habinek is an Assistant Professor of Literature at Bard College. Her work has appeared in the journals *Shakespeare* and *Configurations* and in the literary review *Open Letters Monthly*. She has held fellowships at the Folger Shakespeare and Huntington libraries, as well as with the Institute for Historical Research at the University of London. She is currently working on a project titled *Such Wondrous Science: Brain and Metaphor in Early Modern English Literature.*

David Hawkes is Professor of English at Arizona State University. He is the author of *Ideology: The New Critical Idiom* (1996), *Idols of the Marketplace: Idolatry and Commodity Fetishism in English Literature 1580–1680* (2001), *The Faust Myth: Religion and the Rise of Representation* (2007), *John Milton: A Hero of Our Time* (2010), and *The Culture of Usury in Renaissance England* (2010). Hawkes received a long-term fellowship from the National Endowment for the Humanities, and his work has appeared in a range of academic and popular journals, including *The Nation, Times Literary Supplement, Journal of the History of Ideas, Studies in English Literature, English Literary Renaissance, Milton Studies*, and *Shakespeare Quarterly.*

David Hillman is a Senior Lecturer in English at the University of Cambridge and a Fellow of King's College. He is the author of *Shakespeare's Entrails: Belief, Scepticism and the Interior of the Body* (2007), and the coeditor of several books, including *The Body in Parts: Fantasies of Corporeality in Early Modern Europe* (with Carla Mazzio, 1997), *The Book of Interruptions* (with Adam Phillips, 2007), and *Marx and Freud: Great Shakespeareans* (with Jean E. Howard and Crystal Bartolovich, 2012). He is currently working on a book on greetings and partings in early modern England. Together with Carla Mazzio, David was awarded the English Association Beatrice White Book Prize of 1999 for *The Body in Parts.*

Tiffany Hoffman is an instructor and Ph.D. candidate at McGill University in Montreal, Canada. She is currently completing her dissertation on the topic of shyness in Shakespeare's plays.

Laurie Johnson is Associate Professor of English and Cultural Studies at the University of Southern Queensland, Australia, and current Vice-President of the Australian and New Zealand Shakespeare Association. He is the author of *The Wolf Man's Burden* (2001), *The Tain of Hamlet* (2013), and more than twenty articles and book chapters on cultural theory, early modern studies, ethics, phenomenology, psychoanalysis, Shakespeare studies, and related topics. He is also editor (with Darryl Chalk) of *Rapt in Secret Studies: Emerging Shakespeares* (2010) and past general editor of the Cultural Studies Association of Australia.

Ros King is Professor of English Studies and Head of English at the University of Southampton and a Fellow of the English Association. A musician and theatre director as well as an academic, she has worked as a dramaturg with Shakespeare's Globe in London, the English Shakespeare Company, and Shakespeare Santa Cruz (California), and has extensive experience giving workshops and talks to school students and teachers. She is the author of numerous books, editions, and articles concerning early modern drama, including *The Works of Richard Edwards: Politics, Poetry and Performance in Sixteenth-Century England* (2001); *Cymbeline: Constructions of Britain* (2005); and *Shakespeare: A Beginner's Guide* (2011).

James A. Knapp is Associate Professor of English and Edward L. Surtz, S. J. Professor at Loyola University Chicago. He is the author of *Illustrating the Past in Early Modern England* (2003) and *Image Ethics in Shakespeare and Spenser* (2011). His work has appeared in *Shakespeare Quarterly*, *ELH*, *Criticism*, and *Poetics Today* as well as a variety of essay collections.

Carla Mazzio is Associate Professor of English at the State University of New York, Buffalo, is the author of *Book Use, Book Theory: 1500–1700* (with Bradin Cormack, 2005), and *The Inarticulate Renaissance: Language Trouble in an Age of Eloquence* (2009), for which she was awarded the 2010 Roland H. Bainton Prize for Literature, and editor of several books including *The Body in Parts: Fantasies of Corporeality in Early Modern Europe* (with David Hillman, 1997), for which she was a corecipient of the English Association Beatrice White Book Prize of 1999, and *Historicism, Psychoanalysis, and Early Modern Culture* (with Douglas Trevor, 2000).

Jennifer Rae McDermott teaches Shakespeare and Renaissance Literature at John Abbott College. Her recently completed dissertation at the University of Toronto, "Shakespeare in Another Sense," examines the interplay of senses, stage, and text from the perspective of historical phenomenology in order to probe the nature of audience response. In addition to developing a monograph on Shakespearean sensations, her publications include "Perceiving Shakespeare" for the embodiment issue of *Early Modern Literary Studies* (2009) and "The Melodie of Heaven: Sermonizing the Open Ear" in *Religion and the Senses in Early Modern Europe* (Leiden, Netherlands: Brill, 2013).

Gail Kern Paster is the Director Emerita of the Folger Shakespeare Library and Editor of *Shakespeare Quarterly*. Her most recent books are *Humoring the Body: Emotions on the Shakespearean Stage* (2004) and, with Mary Floyd-Wilson and Katherine Rowe, the collection of essays *Reading the Early Modern Passions: A Cultural History of the Emotions* (2004). She continues to write scholarly essays on the ecology of the passions in the early modern period.

Jan Purnis is an Assistant Professor of English at Campion College at the University of Regina. She has published on the stomach's role in early modern emotion theory and the gendered stomach in *The Taming of the Shrew*. She is currently completing a book manuscript called "Digestive Tracts: Early Modern Discourses of Digestion."

Katherine Rowe is Professor of English at Bryn Mawr, is the author of *Dead Hands: Fictions of Agency, Renaissance to Modern* (1999) and coeditor of *Reading the Early Modern Passions: Essays in the Cultural History of Emotion* (2007). A recipient of grants from the National Endowment for the Humanities, the Mellon Foundation, and the Pennsylvania Department of Education that support her work in the digital humanities, Professor Rowe is Associate Editor of The Cambridge World Shakespeare and a Trustee of the Shakespeare Association of America. She cofounded Luminary Digital Media, a small publisher of next-generation mobile reading experiences, including Shakespeare's *The Tempest* for iPad.

Michael Schoenfeldt is the John Knott Professor of English and Chair of the Department of English at the University of Michigan, where he has taught since 1985. He is the author of *Prayer and Power: George Herbert and Renaissance Courtship* (University of Chicago Press, 1991), *Bodies and Selves in Early Modern England: Physiology and Inwardness in Spenser, Shakespeare, Herbert, and Milton* (Cambridge, 1999) and *The Cambridge Introduction to Shakespeare's Poetry* (2010); and editor of the *Blackwell Companion to Shakespeare's Sonnets* (2006). He is currently editing *John Donne in Context* for Cambridge and working on a book for Blackwell's entitled *Reading Seventeenth-Century Poetry*.

Garret A. Sullivan, Jr., is Professor of English at Penn State, and author of *The Drama of Landscape: Land, Property, and Social Relations on the Early Modern Stage* (1998), *Memory and Forgetting in English Renaissance Drama: Shakespeare, Marlowe, Webster* (2005), and *Sleep, Romance and Human Embodiment: Vitality from Spenser to Milton* (2012).

He has also edited numerous works, among which are *Environment and Embodiment in Early Modern England* (with Mary Floyd-Wilson, 2007), *Early Modern English Drama: A Critical Companion* (with Patrick Cheney and Andrew Hadfield, 2007), *The Cambridge Companion to English Renaissance Tragedy* (with Emma Smith, 2010), and the *Encyclopedia of English Renaissance Literature* (general coeditor with Alan Stewart, 2012).

John Sutton is Professor of Cognitive Science at Macquarie University in Sydney, where he was previously Head of the Department of Philosophy. He is author of *Philosophy and Memory Traces: Descartes to Connectionism* (Cambridge, 1998), and coeditor of *Descartes' Natural Philosophy* (Routledge, 2002) and of the international journal *Memory Studies*.

Evelyn Tribble is the Donald Collie Chair at the University of Otago, Dunedin, New Zealand. Her most recent books are *Cognition in the Globe: Attention and Memory in Shakespeare's Theatre* (Palgrave, 2011), and *Cognitive Ecologies in Early Modern England* (with Nicholas Keene). She has also published *Margins and Marginality: The Printed Page in Early Modern England* (Virginia, 1993), as well as numerous articles on Shakespeare. She is currently at work on a project on Shakespeare, skill, and embodiment.

Index

For Product Safety Concerns and Information please contact our EU
representative GPSR@taylorandfrancis.com
Taylor & Francis Verlag GmbH, Kaufingerstraße 24, 80331 München, Germany

www.ingramcontent.com/pod-product-compliance
Ingram Content Group UK Ltd.
Pitfield, Milton Keynes, MK11 3LW, UK
UKHW022117190926
475702UK00013B/1492

* 9 7 8 1 0 3 2 2 4 2 9 2 7 *